Khaled Chawraba
Abdallah Nasserddine
Hareth Sleiman

Development of polyurethane/carbon nanotube composite materials

Khaled Chawraba
Abdallah Nasserddine
Hareth Sleiman

Development of polyurethane/carbon nanotube composite materials

ScienciaScripts

Imprint

Any brand names and product names mentioned in this book are subject to trademark, brand or patent protection and are trademarks or registered trademarks of their respective holders. The use of brand names, product names, common names, trade names, product descriptions etc. even without a particular marking in this work is in no way to be construed to mean that such names may be regarded as unrestricted in respect of trademark and brand protection legislation and could thus be used by anyone.

Cover image: www.ingimage.com

This book is a translation from the original published under ISBN 978-613-8-43636-2.

Publisher:
Sciencia Scripts
is a trademark of
Dodo Books Indian Ocean Ltd. and OmniScriptum S.R.L publishing group

120 High Road, East Finchley, London, N2 9ED, United Kingdom
Str. Armeneasca 28/1, office 1, Chisinau MD-2012, Republic of Moldova, Europe
Printed at: see last page
ISBN: 978-620-6-22157-9

TABLE OF CONTENTS

Thanks

First of all, we would like to thank Dr Hares SLEIMAN, who supervised this work throughout the year, and led all the meetings and stages of the project. We are deeply grateful to him for having placed his trust in us from the very start of our Master's program, and for having accompanied us during these nine months to bring this work to a successful conclusion. Thanks to his encouragement, and at times his scientific firmness, we learned to overcome our fears and adopt an appropriate work methodology to achieve promising results.

We really appreciated Dr wassef el khatib's scientific and human complementarity, which enabled us to work together in the same laboratory under excellent conditions on an exciting subject, as a member of the project monitoring committee and finally as a member of the jury.

We would like to thank Layla ghanam for agreeing to report this thesis manuscript.

We would also like to thank the industrial COMPANY -Api- (Jeita- Lebanon), especially its CEO Mr. Antoine CHEDID, for welcoming us and providing us with materials and equipment. We would also like to express our sincere thanks to the specialists who helped us carry out the mechanical properties tests, in particular: -Mr. Richard Issa: Technical quality control manager, Mr. Jean Kayrouz: Lab Supervisor, and -Mr Rabih Kassas : Lab supervisor

We would like to thank Amina Mourtada for his interest in our work and his help with the homogeneity test experiments.

We would like to thank Dr. Roland Habchi for carrying out the SEM experiments in the platform laboratories at F.Sc-2-Fanar, and Selena khayrallah for successively performing the TGA measurements,

We would also like to *thank* LUBNA BITAR for her advice, availability and sympathy.

INTRODUCTION

Polyurethane (PU) is one of the most interesting synthetic elastomers[2] . Due to its unique properties; PU's linear segmented structure appears in the form of (A-B)n . The flexible B segment is normally formed from a macromolecular polyester or polyether with a molecular weight of between 1000 and 3000, while the hard A segment consists of a di-isocyanate elongated by reaction with a low-molecular-weight diol or diamine. Because of the difference in chemical structure, the microdomains forming the soft and hard segments are joined together by mutual attraction resulting from intermolecular hydrogen bonds.

Since their discovery in 1991 by Iijima,[NEW] the unique and novel properties of carbon nanotubes (CNTs)[19] and the technological possibilities offered by CNT-polymer composites, such as improved mechanical, electrical and thermal properties, continue to attract worldwide research attention. The literature to date has converged on methods for distributing and dispersing carbon nanotubes within the polymer matrix, and on how interactions between polymer chains and carbon nanotubes can be promoted.

Thus, the objectives of this thesis work were firstly to prepare new polyurethane-carbon nanotube composites functionalized with two types of functional groups, one of which increases nanotube-polymer chain interactions, the other of which decreases these interactions, and secondly to study the variations in thermal stability measured by

A.T.G. analysis and elongations at break examined by a tensile tester.

The samples examined belong to six families, each grouping PU-NTCs with different levels of carbon nanotube fillers.

This brief is divided into four parts;

- The first chapter consists of a general presentation of the bibliography concerning the different methods of dispersion and incorporation of the

various types of carbon nanotubes in the polymer matrix, as well as the different methods of functionalization, especially the Carboxylation and acylation of carbon nanotubes.

We then reviewed the various methods of synthesizing and processing PU-NTCs, resulting in either a film, a fiber, a thermoplastic foam or a shape-memory polymer.

The aim of this chapter is to place the work in context, and in particular to highlight the influence of CNTs on the mechanical and thermal properties of polyurethanes.

- The second chapter shows a summary of six families of composites, each family comprising 4 to 6 samples, and details the quantities of reagents used in the synthesis of each sample.

- The third chapter examines the hardness and homogeneity of materials as a function of CNT loading, and presents micrographs of samples obtained by scanning electron microscopy. The results of thermal and mechanical analyses, such as thermogravimetric analysis and tensile testing, are also discussed and exploited,

The fourth chapter presents the materials, reagents and protocols for functionalizing CNTs, as well as the experimental techniques used.

Chapter 1. POLYURETHANE COMPOSITES - CARBON NANOTUBES; SYNTHESIS, PROCESSING, AND MECHANICAL AND THERMAL PROPERTIES

I. Incorporating carbon nanotubes into the polymer matrix

1. TYPES OF CARBON NANOTUBES (NTC)

Two main types of carbon nanotube are commonly used [1.. p=]: multi-walled nanotubes (MWNTs), which are composed of concentric tubes of different radii, and single-walled nanotubes (SWNTs). Carbon nanotube diameters range from 1.4 to 100 nm for MWNTs and from 0.4 to 3 nm for SWNTs, with aspect ratios of 1000 or more. Moduli for MWNTs and SWNTs are approximately 270 GPa and 1 TPa respectively, while tensile strengths are 11 and 200 GPa, respectively. In fact, the electrical properties of single-walled nanotubes (SWNTs) are much better than those of multi-walled nanotubes, and the inner walls of (MWNTs) do not contribute significantly to the mechanical properties of the material[6] . Single-walled carbon nanotubes, in addition to their superb mechanical properties, also have excellent electrical properties, and a conductivity 5 times greater than that of copper. The thermal conductivity of (SWNTs) is also extremely high, and greater than that of diamond or copper, such properties suggesting that these nanotubes should be excellent materials for incorporation with polymers[6] .

Carbon nanotubes still in use[20] , which are vapor-grown carbon fibers (VGCFs) (Showa DenkoK.K.), **have a** diameter of around 150 nm and a width of 10-20 μm.

2. DIFFERENT METHODS FOR DISPERSING (NTC) IN POLYMER

Dispersing nanotubes in the polymer was always a key issue. In polyurethane manufacturing, the three main routes for dispersing nanotubes in a polymer are[1] :

a.   Melted mixture.

b.   Dispersion of nanotubes in a solvent and dissolution of polymer in the same solvent, followed by evaporation of the solvent.

c.   A reaction (polycondensation) of monomers or prepolymers *(polyurethane thermoplastics are made by a two-stage procedure)* in the presence of dispersed nanotubes.

Most often, CNTs can be dispersed using an ultrasound probe[1] [3][23] , a surfactant to stabilize the emulsion [3,25] or a volatile solvent to reduce the viscosity of the mixture of reagents and CNTs. [23]

Carbon nanotubes ₁.₋15p44 2l are strongly affected by vander Waal's forces, which would give rise to the formation of aggregates and make it difficult to disperse CNTs in the polymer. The challenge in developing a high-performance polymer-CNT composite is the homogeneous dispersion of CNTs in the polymer matrix.

D.X. Yan,... W.Zhang & all [23] have fabricated a low-percolation conductor formed from a PU composite foam with CNTs dispersed simultaneously in the cell walls and in the PU foam spacers, using ethanol as a diluting agent, which reduces the viscosity of the polyether polyol, and optimizes the ultrasound-assisted dispersion of carbon nanotubes ... . Homogeneous distribution of carbon nanotubes and significant charge transfer between the carbon nanotubes and the PU matrix produced a 31% improvement in compression properties, and a 50% increase in static modulus at room temperature. Thermogravimetric analysis (TGA) shows that the introduction of a low CNT content induces remarkable thermal stabilization of the[23] matrix.

3.   MORE EFFICIENT INCORPORATION OF NTCS INTO THE MATRIX

The use of a composite with excellent properties to prepare a high-performance polymer was prevented by the poor dispersion of CNTs due to the lack of chemical compatibility between the two phases[3] . Chemical functionalization of CNTs has proved effective in improving both the dispersion and compatibility of nanotubes and matrix.

The modification of polymer-embedded carbon nanotubes can be divided into two categories, involving either non-covalent bonding or covalent bonding between CNTs and polymer. Non-covalent modification of CNTs involves the physical adsorption or packaging of polymers on the CNT surface. The advantage of this modification is that it does not destroy the CNT conjugate system and does not influence the final structural properties of the materials. The second modification is the chemical covalent bonding (grafting) of polymer chains to CNTs, where strong chemical bonds between nanotubes and polymers are established.

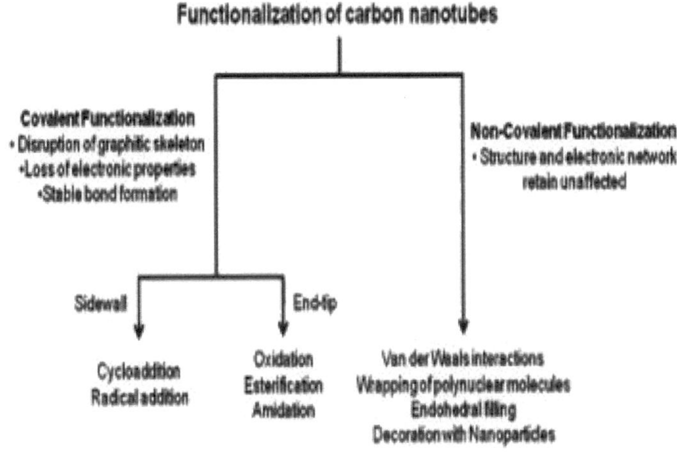

Figure 1- Explanatory diagram of carbon nanotube functionalizations [29]

4. CARBOXYLATION AND ACYLATION OF CARBON NANOTUBES

Generally, modified CNTs are prepared in two stages. In the 1ère step, carboxylic acid groups can be introduced into the sites of surface defects during and after treatment of CNTs with an oxidizing agent, and then these groups can be further converted into other active groups, such as an amide or ester moiety, which can react directly with the polymer or monomers.

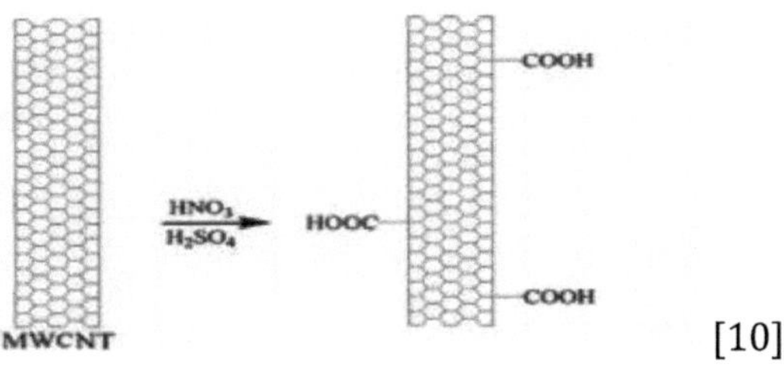

[10]

CNTs with oxidant-generated carboxylic acid groups are often used as precursor compounds for polymer-CNT composites.

Chemical treatment to functionalize MWNTs is produced by several routes; the first is achieved by a mixture of concentrated nitric and sulfuric acids[3] [9] [16] [18] , in a molar ratio of 1: 3, respectively.

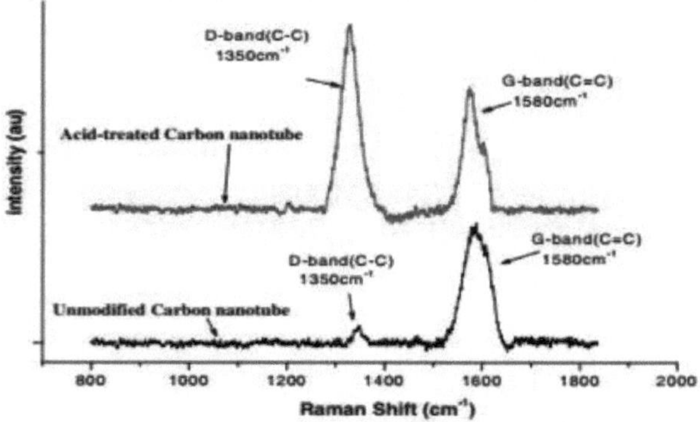

Fig. 1. Raman spectrum of the pristine carbon nanotube and the modified carbon nanotube.

Figure 2 [2] -Raman spectrum of carbon nanotubes and modified nanotubes

In a typical experiment, 1 g of crude carbon nanotubes were added to 40 cm³ of the acidic mixture in a flask, and heated to reflux for 20 min. On cooling, the mixture was washed with distilled water until the wash water showed no acidity.

The other is a selective oxidation[29] of SWCNT by hydrogen peroxide was carried out, varying heating times and monitored by ultraviolet-visible and near-infrared spectroscopy. A significant increase in relative absorption intensity indicated a content of over 80% metallic SWCNTs in the final product. Semiconducting SWCNTs were found to be more reactive than metallic ones due to the hole-doping effect of hydrogen peroxide, resulting in faster oxidation. Recently, the controlled shortening and uncorking of very thin MWCNTs has been achieved using a mild oxidizing agent (H_2O_2 in 15% aqueous solution) and reaction conditions (100°C for 3 h). [29]

Oxidized MWCNTs can also be prepared[25] in a medium containing 300 cm³ of a 0.5M H_2SO_4 solution, 1.5 g $KMnO_4$ and 0.5 g MWCNTs. The dispersion is sonicated at 40°C for one hour. The product is filtered and washed with concentrated HCl to remove MnO_2.

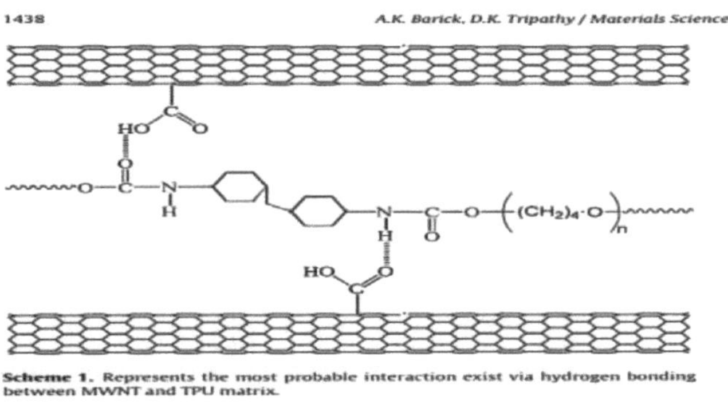

Scheme 1. Represents the most probable interaction exist via hydrogen bonding between MWNT and TPU matrix.

Figure 3-Representation of likely interactions between MWNT nanotubes and TPU matrix [12]

In general, two methods have been used to fabricate polymer-CNT composites: direct incorporation of functionalized CNTs into the polymer matrix (a non-covalent attachment) and surface in situ polymerization (a covalent attachment)[10].

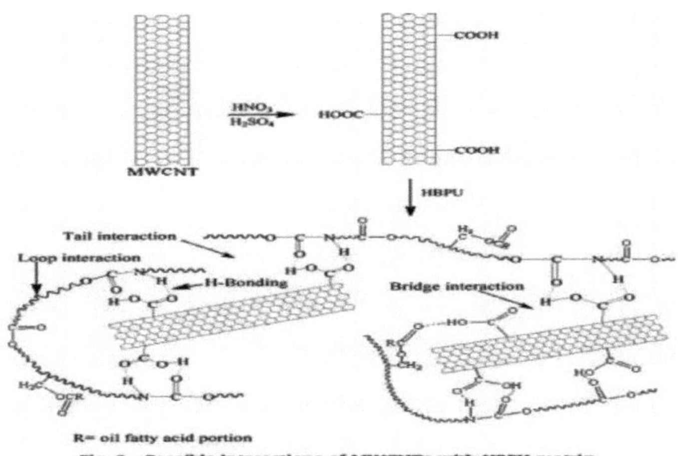

Fig. 6 – Possible interactions of MWCNTs with HBPU matrix.

Figure 4-Representation of likely interactions between MWNT nanotubes and TPU matrix [10]

Scheme 1. Covalent bonding and ionic bonding between the carbon nanotube and waterborne polyurethane.

Figure 5 - Covalent and ionic bonds between carbon nanotubes and PU with aqueous ends

Covalent bonds (attachments) are considered to be more advantageous than non-covalent attachments for preparing polymer-CNT composites with excellent properties.

J.Xiong et al[3] have succeeded in transforming the COOH carried by carbon nanotubes (CNTs) into the more reactive COCl acyl chloride function in a Thionyl chloride dispersion subjected to ultrasonic frequencies. The PU-CNTs composites are obtained by reaction with ethylene diamine followed by the addition of polyol and di-isocyanates (see IR Spectrum).

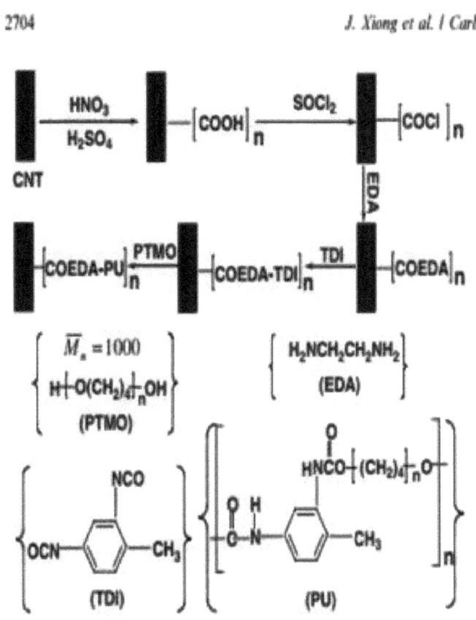

Fig. 2. Chemical routes for preparation of PU–CNT composite.

Figure 6-The chemical route for preparing a PU-CNT composite

44 (2006) 2701–2707

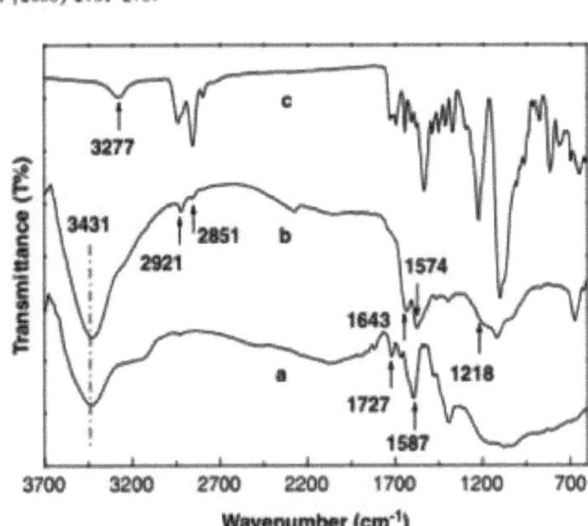

Fig. 4. FTIR spectra of (a) acid-treated CNT (in a KBr wafer), (b) CNT amide derivative (in a KBr wafer) and (c) PU–CNTcomposite.

Figure 7- Spectra of a)Acid-treated CNT (KBr pellet) b) Amide-derivatized CNT (KBr pellet) c) PU-CNT composite

5.   DIFFERENT METHODS OF OPERATING (NTC)

Y. Bai & all[14] have prepared a series of shape memory polymers (SMPC) with different carbon nanotube contents (2.4, 3, 4, 5, 10, 12%). This is a multi-walled polyurethane-carbon nanotube composite (MWCNT) with exceptional properties obtained by in situ polymerization. This material (SMPC) is prepared as follows: (1) the pre-polymer will be obtained by a ring-opening polymerization of propiocaprolactone ε-PCL; the polymerization was initiated by reactions with hydroxyl groups attached to the surfaces of MWCNT multiwall nanotubes. Dispersion of MWCNT-OH is achieved by ultrasonic waves for 30minutes, its stability of is a crucial parameter for the performance of the composite material, the authors found that MWCNT-OH nanotubes are uniformly dispersed in ε-CL after one month, while these nanotubes precipitate just after one day in DMF.An amount of 5% by mass of tin octanoate (catalyst) will be added after dispersion, in a second step, (2) isocyanates will be used to bind and bridge the pre-polymers and obtain the shape memory polyurethane composite (SMPC}.

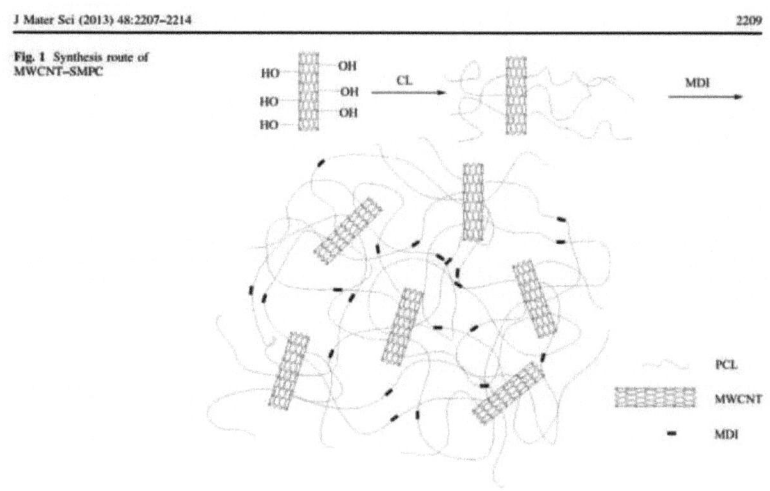

Figure 8- **The MWCNT-SMPC synthesis route**

Generally, the degree of carboxylation is relatively low with ordinary treatment of CNTs with a strong oxidant. A group of researchers at[17] has used a convenient route to functionalize carbon nanotubes by radical addition reaction of a free radical formed by an alkyl group and terminated on the other side by a carboxylic acid (Figure 9).

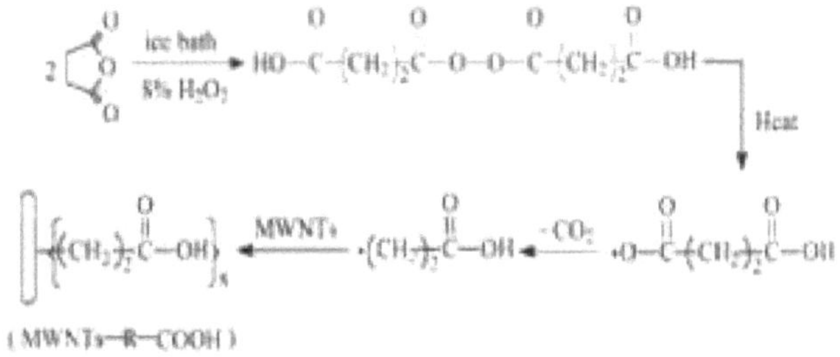

Scheme 1 Preparation of sidewall acid-functionalized MWNTs.

[17]

Figure 10-   **Preparation of acid-functionalized sidewall MWNTs**

This method is scalable and provides a pathway to multi-walled carbon nanotubes, functionalized by groups that will be adapted for further elaboration.

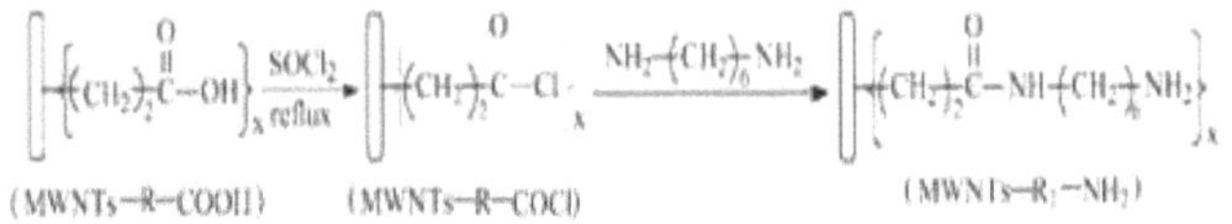

Scheme 2 Preparation of amido derivatives of MWNTs.

[17]

Figure 11-   **Preparation of MWNTS amide derivatives**

The aim of this step is to add COOH. Subsequently, the nanotubes

Amino-functionalized MWNTs (MWNT-R1-NH2) will be prepared as shown in Figure11.

Functionalization[17] * * * [21] of carbon nanotubes is also very effective, by preparing soluble carbon nanotubes, generating structural defects with a strong acid and inducing a chemical reaction between alkyl groups and azide moieties, and a π-π interaction with electron-rich molecules.

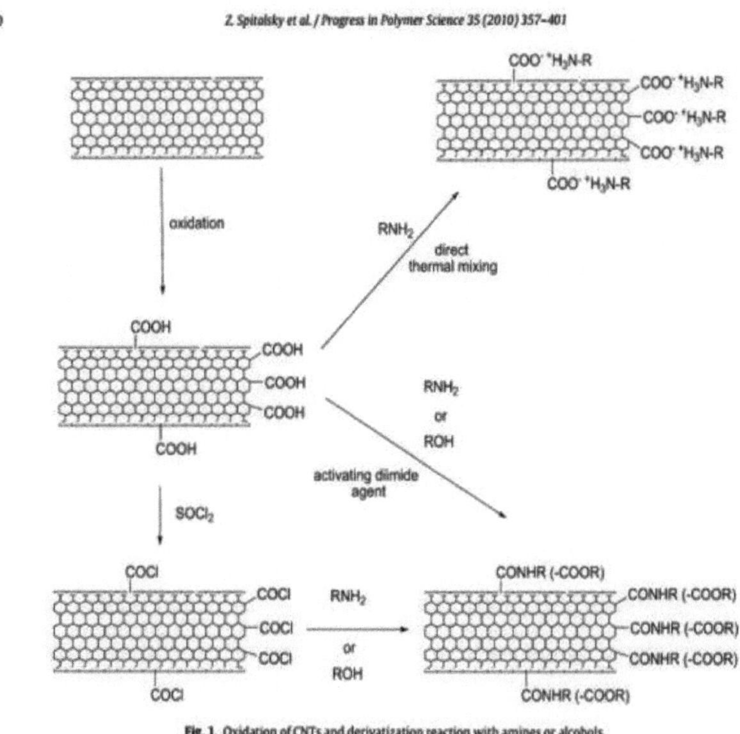

[24]

Figure 12- Oxidation of carbon nanotubes and their reaction by-products with amines and alcohols

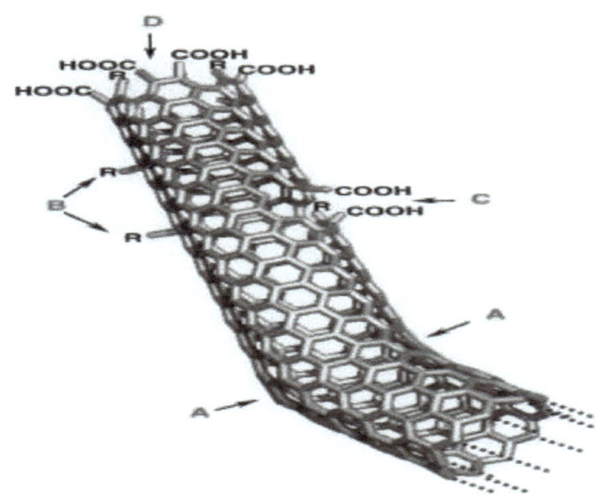

Figure *13-Nanotube end and wall defects.*

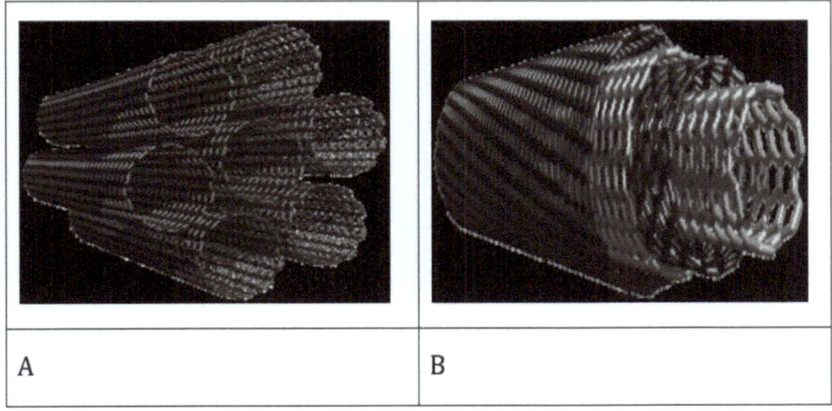

| A | B |

Figure *14-Schematic* representation of *the two nanotube classes single-sheet carbon nanotube (SWCNT)* (A) *and multi-sheet carbon nanotube (MWCNT)* (B)

The esterification reaction was used[24] for grafting polyethylene glycol (PEG) to acyl chloride-activated SWCNT chains in the absence of a solvent in the medium. The reaction, which grafted PEG possessing terminal hydroxyls to thionyl chloride-treated MWCNT nanotubes, was carried out at temperatures above the polymer's melting point. However, grafting efficiency was low, with TGA analysis showing the presence of 93% by weight of nanotubes.

The two oxidized nanotubes SWCNT and MWCNT were the subjects [24] of carbo-di-imide-activated esterification with a polyimide derivative bearing alkoxysilane groups at the ends or having lateral hydroxyl groups. From integrations of 1H NMR signals, the CNT content in the sample was estimated to be around 35% by weight.

A similar approach was used in the case of a poly (N- vinylcarbazole) copolymer containing lateral hydroxyl groups , which were grafted to SWCNT oxidized nanotubes under conditions typical of an acyl-activated esterification reaction . CNT content in the sample was estimated at ~20wt%.

Wang and T. seng prepared block polyurethanes, the reagent they used, as a chain extender, carries carboxyl groups. This polymer undergoes an esterification reaction with MWCNTs activated by acyl chloride. According to TGA data, CNT content can be as high as 67% by weight. It was found that a longer acid treatment time of the CNT material resulted in a higher amount of grafted polymer chains [24].

In addition, several types of grafting have been encountered, such as grafting by radical mechanism, nucleophilic addition, cycloaddition, condensation, atom transfer radical polymerization.

II. POLYMERIZATION AND SYNTHESIS :

to obtain a material with both the advantages and properties of the PU matrix and the thermomechanical and electrical properties of carbon nanotubes, various methods[21] have been examined in efforts to solve this problem, such as mechanical blending, chemical functionalization of nanotubes, and the use of molecular and macromolecular organic surfactants through non-covalent interactions. In situ polymerization, where nanotubes are incorporated into the prepolymer during the polymerization process, has been used to provide reinforced carbon nanotube composites with improved mechanical properties.

Broadly speaking, there are two approaches[5] : the first involves modifying the molecular structure of polyurethane by altering its three basic components: a polyol, a diisocyanate and the chain extender. The type of polyol plays a very important role in polyurethane properties. Compared with polycaprolactone polyol (PCL), a polyether polyol such as polypropylene glycol (PPG) has low tensile strength, low elongation and low thermal stability, but it's low cost. The second is to introduce the inorganic filler into the polyurethane matrix. The disadvantages are that the addition of these inorganic fillers often worsens behavior under stress and reduces elongation at break.

Several methods have been developed for the synthesis of carbon nanotube-polymer composites[15] : **in situ polymerization of monomers,** in which the carbon nanotubes are added to the pre-polymers during the polymerization process, while for **compression molding from a polymer melt,** the CNTs are mechanically dispersed in a polymer melt at high temperature and with the aid of a mixer. In **solution molding,** a solvent is used to dissolve a polymer, and the carbon nanotubes are added to the solution. Composite films are molded from mixed solutions.

1. In situ polymerization of monomers

Preparation of a stable polyol-CNT dispersion can be carried out either in the presence of a surfactant [5], using ultrasound frequencies [7. .1816] or non[1 5,6] , or by dissolving the polyol in an organic solvent (DMF) and then emulsifying the solution with water, with the addition of a water dispersion of CNTS.[15,18,22] .

In situ polymerization of monomers in the presence of CNT[16,6] and PCL2000 was first brought to reaction with MDI at 80⁰ C for 2 hours to produce the prepolymer. Subsequently, a chain extender, 1,4-Butanediol (BD), was gradually added to the prepolymer at 110°C for one hour. After washing with water, the PU was completely dried in a vacuum oven. In this study, a molar ratio of 5: 1: 4 for MDI / PCL / BD was used to obtain 44.6% percent hard segments.

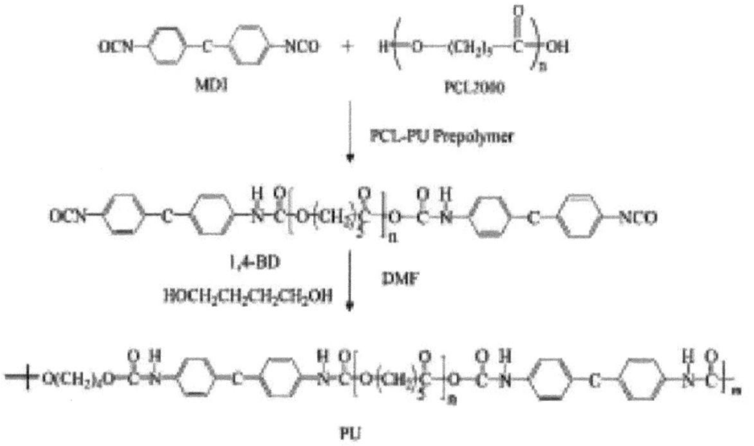

Scheme 1   The two-step polymerization process of polyurethane.

[6,16]

Figure 15- Two-step polyurethane polymerization

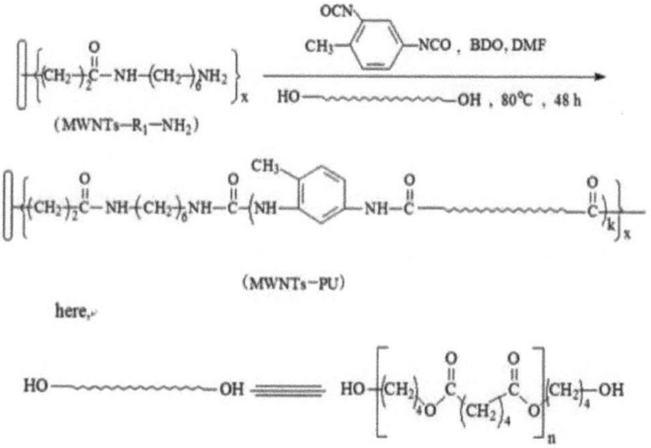

[17]

Figure 16-Polycondensation for the preparation of functionalized PU-MWNTs

Researchers[21] have carried out a comparative study of the dispersion structure

of carbon nanotubes, using three different polymerization processes; namely;

- A classic blend,

- In situ polymerization

- Cross-linking polymerization

Optimum dispersion of nanotubes in the polyurethane matrix is achieved by using carbon nanotubes chemically modified by attaching carboxyl groups to their walls, using the bridging or cross-linking polymerization method. The hybrid structure of the nanotube/polyurethane covalent bonds effectively prevents reaggregation of the nanotubes in the polymer matrix].[21

CARBON 48 (2010) 1598–1603   **1601**

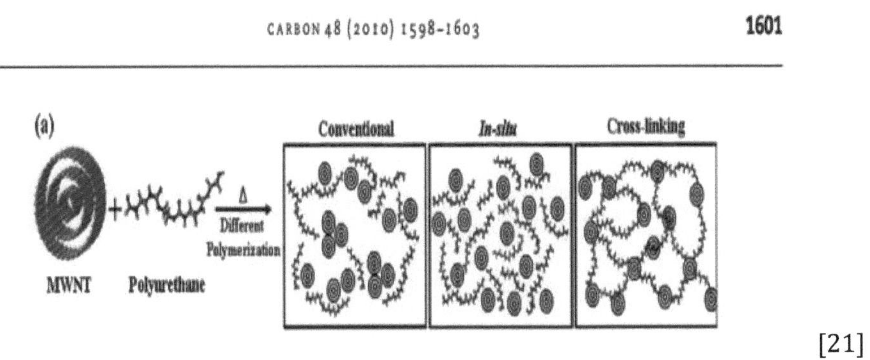

[21]

Figure 17- Polymerization processes for PU shaping

a) Films

Three methods were used[15] to prepare films of PU-MWNTs nanocomposites; Method (1) the prepolymer is prepared from a reaction of MDI and PCL. Following the chain extension step, a calculated amount of MWNT and BD was added to the pre-polyurethane solution.

Method (2): the required weight fractions of MWNTs were first dispersed in Polycapro lactone (PCL). MDI was then added to this mixture. BD chain extender was added to this pre-polymer, and the final composite PU MWNT was synthesized. In both cases, the composite films were prepared in a hot press at 140^0 C for 10 minutes.

Method (3): PU was first dissolved in dimethylformamide (DMF), a solution at room temperature. MWNTs were then added to this solution and stirred for 24 h. In all these cases, the amount of MWNT is 3% by weight relative to the polyurethane[15] .

b)   Memory fiber

Shape memory polyurethanes were synthesized[6,13,14,16,21] with (PCL) polycaprolactonediol as the flexible segment, while 4,4- 0-diphenyl methane diisocyanate (MDI) and 1,4-butanediol molecular extender (BDO) were used as the hard segment.

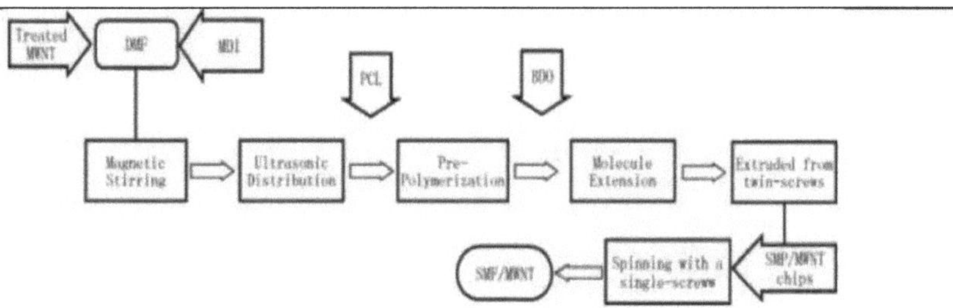

Figure 18-Polymer synthesis and fiber preparation process

The polymer synthesis process and the fiber preparation process are shown in both figures (14) and (17). A calculated amount of SWNTS[6] or MWNTs[16] modified by an oxidizing agent or carrying a Hydroxyl group [CNTsOH][14] was pre-distributed in DMF . While the MDI melt is maintained at 80°C, and was subjected to vigorous stirring for half an hour and ultrasonic vibration for 1 h. Shape memory polyurethane was prepared by in situ polymerization. BDO was added with vigorous stirring. The entire reaction was carried out in a high-purity nitrogen environment.

PU-CNT SMP nano-composites[20] have been developed with typical carbon nanotubes, which are vapor phase growth fibers (VGCF). A fine, homogeneous dispersion of VGCF through the SMP matrix is achieved.

Polyurethane-MWNTs composites, synthesized[15] by in situ polymerization, had better properties than those prepared by solution molding. MWNTs acted as nucleating agents and, as a result, significantly increased the crystallization temperature of PU. The original shape of the sample was almost recovered in 40 seconds with the bending mode when a 50 V electric field was applied.

Processing can be carried out by several techniques, the composite films will be prepared under a heated press[15] at 140⁰ C for 10 minutes, while the fibers and filaments are obtained by extrusion; The reaction mixture was injected into a twin-screw extruder[6,16] , and SMP-MWNT specimens were then produced. The polyurethane obtained had an average molecular weight of 1.37 x 10^5 . 40-denier SMP monofilaments were spun by a 20 mm single-screw extruder. Temperatures in the first zone, second zone, third zone, fourth zone, extruder head, spin pack, melt pipe, and pump were 180, 205, 208, 210, 212, 212, 212, and 212⁰ C, respectively. Laminar air temperature was 22C. Extruder head pressure was 5.00 MPa, and rotational pressure was 22.00 MPa. A die with a single 0.4 mm hole was used. The fiber was then passed over two pairs of rollers at 80°C to release internal stress. Winding speed was 100 m/min.

Shape memory polymers (SMPS) have been under intensive investigation in recent years. They can be deformed and fixed in a temporary shape. They can also recover their permanent shape by responding to various external stimuli, e.g. heat, light, electricity, magnetic field... etc. PU is a kind of elastic thermoplastic made up of flexible and hard segments. These elastic materials appear in two distinct phases. Its shape-memory effect is generally induced by thermal stimulation, which occurs by heating flexible polyurethane segments above their transition temperature (glass transition temperature or melting temperature). The flexible segments, on the other hand, act as molecular switches and are responsible for recovering the permanent shape from deformed temporary forms.

The temperature at which the shape memory effect is interrupted varies from 44 to 55⁰ C, depending on the flexible segment content, which ranges from 50 to 90% by weight. The molecular weight of the flexible segment must be low when used at low temperatures.

The electroactive [hybrid PU-SWNCT] shape memory effect was induced by the release of thermal energy. In this case, the hybrid PU- SWCNT would act as an electrical resistor. When voltage has been applied, current flows through the resistor, and the power consumed generates thermal energy. The temperature of the resistor surface then rises. The electroactive shape memory became more active as the induced thermal energy exceeded the melting point of the flexible segment of hybrid PU-SWCNTs. The optimum condition for conductive networks in PU-SWCNT hybrid materials was found to be 3% SWCNT incorporated in PU. PU-SWCNT hybrids containing 4% SWCNTs showed good electroactive shape recovery properties when an electrical voltage was applied at low temperature. The original shape of the sample was about 88% recovered in 90 s when 30 V, 0.1 Ampere power was triggered. The recovery[13] of shape by conductive PU composites, based on carbon nanotubes, was invested by applying an electrical voltage, and not by applying thermal heating. This performance may lead to the application of shape memory polymers as electroactive actuators, which is important in many practical applications, such as intelligent actuators for the control of micro air vehicles[13] .

2. SOLUTION casting method

a)   Foam :

MWCNT nanotubes were functionalized by treatment with hydrogen peroxide (H_{2o2})[22] , a ball mill was used to reduce the length of the MWCNT, a small amount, calculated according to the sample, of MWCNT is added to the polyol solution. Next, distilled water, catalysts and surfactant (Poly(siloxane ether)) were added to the polyol solution and stirred with a mechanical stirrer until the mixture became a homogeneous phase. Cyclopentane was added to the last polyol mixture and stirred again for 10 s at a stirring speed of 3,000 rpm.

The rigid PuF sample is prepared by mixing both a polyol solution

(Pentaerythritol-based polyether), and the polymer 4,4- diphenylmethane diisocyanate (polymeric MDi) at 25°C using a mechanical shaker, with a rotation speed of 5000 rpm for 10 s at room temperature. Finally, the mixture is immediately poured into an open mold (250mm x 250mm x 250mm) to produce freely expanding foams. The mixture is maintained at room temperature for 10 min. The foam is then removed from the open mold and cured at room temperature for at least 1 day before characterization.

b) Movies :

Solution casting is one of the simplest manufacturing processes for surface coating or thin film production Jung et al [8] manufactured a transparent PU film incorporated with functionalized multiwall carbon nanotubes (MWNTs) by solution casting, using vacuum ultraviolet (UV) light treatment. They reported a 2-fold and 10-fold increase in tensile strength and elongation modulus, respectively, for a PU/MWNT composite film compared with a pure PU film.

The manufacture of PU-MWNT, PU-PPy, and PU -MWNT-PPy composite films[18] , was carried out according to the following procedure. The required weight fractions of MWNTs that have been modified with polypyrrole (PPy) are first both dispersed in a DMF solution subjected to ultrasonic frequencies at room temperature for 1 h using a high-power ultrasonic processor. Next, polyurethane was added to this solution, which was stirred for 1 h. The mixtures were then ultrasonicated again for 1 h. Composite films were poured from the solution mixture onto the flat bottom of a clean Petri dish and will be dried completely in an oven. PU- MWNT-PPy composite films are also prepared; in situ polymerization of pyrrole to polypyrrole which will be deposited as a layer on the surface of a PU-MWNT film, and using a chemical oxidation polymerization process[18] .

c) Manufacture of nanofibers and films[7] electrospinning and solution casting

Both techniques; electrospinning and solution molding are simple methods that appear promising for the large-scale manufacture of polymer- MWNT composites in materials engineering applications. Two composite materials; pure PU and PU- MWNTs made by electrospinning and solution molding, can be prepared from the same solution, a comparison of mechanical and thermal properties and their morphologies and structures have been characterized and compared.

A solution of pure high molecular weight thermoplastic polyurethane (PU) was prepared by dissolving, by magnetic stirring, pre-dried pellets (PU 10% by weight) in DMF / MEK (50%-50%, by weight.). MWNTs were dispersed in a bath subjected to ultra-sonic frequencies in DMF / MEK for 3 h, replacing distilled water in the bath every 30 min. The dispersed MWNT solution was added to the pure PU solution, and the mixture was stirred magnetically for 20-24 h. The final weight of MWNTs relative to the weight of PU in the dried web of composite nanofibers and film was 0.1.%.

The Figure 19(a) electrospinning system consists mainly of a high-voltage electrical source, a syringe with a grounded metal needle, a syringe pump, and a grounded flat plate collector, all of which are placed in a closed chamber. As the needle has an internal diameter of 0.51 mm, it was oriented perpendicular to the collector, which is 150 mm from the tip of the grounded needle.

Electrospinning was performed at 11 kV with a feed rate of 1 ml/ 1h. A volume of four ml of prepared pure PU solution, or PU-MWNT / was electrospun, on the flat collector at room temperature and 20-30% relative humidity.

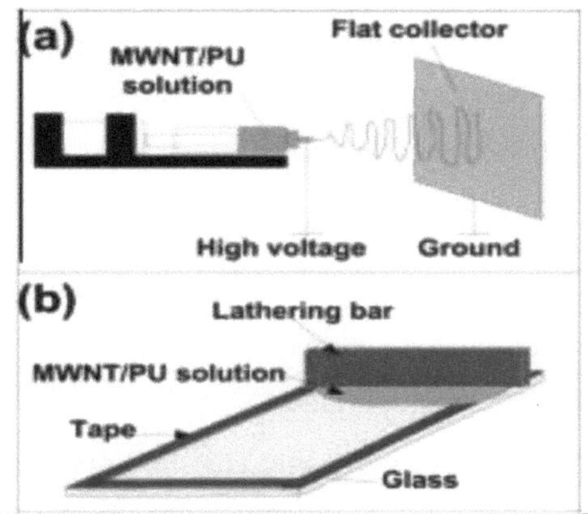

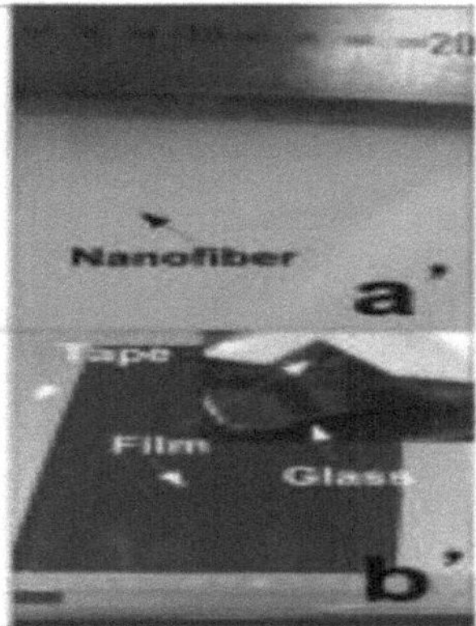

Figure 20-Schematic of electrospinning system (a), solution-molding method (b), images of electrospun nanofibers (a') and solution-molded film (b').

In the solution mold Figure 19 (b), the set-up consisted of a glass substrate, adhesive tapes and a solid metal bar. The pure PU or PU- MWNT solution was poured slowly onto the glass substrate with adhesive tapes, forming a rectangular mold, and spread manually by a metal bar to form a uniform film. Adhesive tapes were used to control film thickness. The solution was poured under ambient conditions. Both the electrospun nanofiber mat and the solution-cast film were dried in an oven at 80^0 C for 48 h to remove residual solvents[7] .

3. MELT-FABRICATED COMPOSITES

The melt-blending process[21] is of particular interest for the manufacture of polymeric nano-composites, because of the ease with which the process can be scaled up to an industrial standard without much hindrance to ecology and economy. The advantage of this technique is its speed and simplicity, not to mention its compatibility with standard industrial processing techniques. The driving force for dispersing CNTs in a polymer matrix is the high processing temperature and high shear force developed during melt processing. Important factors for the production of highly CNT-reinforced polymeric nano-composites using a melt-crosslinking technique, are widely established as (i) an efficient homogeneous dispersion of carbon nanotubes in a polymeric matrix and (ii) the introduction of a strong interfacial interaction between the polymer matrix and the carbon nanotubes, to constitute an efficient interface characteristic of stress transfer.

Shape memory polyurethane fibers can be prepared by wet spinning, dry spinning, chemical spinning and melt spinning. Shape memory fibers (sMF) have outstanding mechanical properties due to their molecular orientation. In addition, they can have many different specific applications. The melt-spinning process does not require harmful di-methyl formamide (DMF) or a coagulation bath, and is relatively economical in terms of production cost[21] .

The sMps used in melt spinning must have greater thermal stability, good rheological properties, and a relatively high molecular weight to obtain shape-memory polyurethane fibers with good performance.

However, the shape-memory polyurethane fibers prepared are generally of much lower tenacity, and the measured shape-memory effect is virtually unremarkable. Multi-walled carbon nanotubes (MWNTs) are excellent candidates for SMF reinforcement and, consequently, for enhancing the shape memory property. In this study, MWNTs are incorporated into a shape-memory polyurethane matrix by in situ polymerization. Through mechanical agitation, ultrasonic vibration, melt mixing, extrusion and spinning processes, a relatively homogeneous distribution of shape-memory polyurethane MWNTs was achieved[21] .

a)   Movies :

The melt-blending method is still used to obtain films of PUT / CNT nano-composites, of different The formulations[12] , which were prepared by mixing MWNT nanotubes with the PUT matrix in an internal batch mixer. Sample mixing was carried out at a temperature of 185^0 C with a rotation speed of 100 rpm and for a time of 8 min. PUT pellets and MWNTs were dried prior to processing to remove water content at 80 and 120^0 C in a preheated vacuum oven for 6 and 24 h, respectively. The compounds were passed once through a cold two-roll mill immediately after batch shuffling to obtain thick sheets that would then be cut into small pieces.

2 mm thick sheet samples of all compositions were prepared using a compression molding machine at 185^0 C for 3 min with a pressure of 5 MPa. [12]

b) Fiber :

Nano-composites with various CNTs loadings were produced[9] with thermoplastic polyurethanes (PUT). Polymer chips and MWNTs powders, which underwent carboxylation treatment with a sulfo-nitric mixture, were dried, weighted and then blended in a mini-laboratory twin-screw extruder. The mixtures were processed for 15 min at 180^0 C and 60 rpm using a counter-rotating screw configuration. **Fiber** samples were then **extruded** through **a** 1.75 mm **cylindrical die**, melt drawn and wound at around 0.25 m / s to produce a final fiber diameter of around 200µm, determined by scanning electron microscopy.

The morphology of the MWNTs and composite fibers were examined by field emission scanning electron microscopy and transmission electron microscopy. SEM images of the fracture surface of the transverse direction of pure PU fiber and PU composite fibers containing 5.6, 9.3, 17.7% by weight of MWNTs, respectively. Well-dispersed points and sharp lines, which are the embedded MWNTs, were observed to compare and increase with CNT loading. It was found that the best MWNTs/in PUR dispersions can be obtained for loading samples up to 9.3% MWNT by weight, while localized concentrations of CNTs were observed for the composite containing a higher concentration of carbon nanotubes, e.g. 17.7% by weight. The high viscosity of the molten polymer-CNT composite limits the concentration of CNTs in the melt-spun PU fibers. From SEM micrographs of the composite fiber samples, prominent dots or lines also suggest that the incorporated MWNTs have been enveloped by the polymer materials. Closer examination of the nanotube reveals a layer, thicker PU, appears to cover the surface of the MWNTs, indicating some degree of wetting and adhesion phase. Most of the MWNTs were observed to be broken rather than simply pulled out of the matrix, upon tearing of the samples, indicating that the interconnection of MWNTs to PU matrix is very strong[9].

c) Shape memory nano-composites with low trigger temperature by melt mixing)

the commercial polyurethane[28] ester-based (PU) with low glass transition temperature was used to elaborate nano shape memory composites (SMP) with low release temperature, 1% by weight of CNT with PU matrix were mixed, the whole was melted at 185 C following the melt mixing method, and subjected to a stirring of 30 rpm for 20 minutes using a mechanical mixer, after having been dried at 80^0 C for 20 h. Samples were prepared using a compression injection machine. Injection temperature and pressure were 205 C and 600 MPa, respectively. The nanocomposites have a

high shape retention and a recovery rate of over 98%. The oxCNT / PU nano-composite shows a higher shape retention rate for the first cycle, faster recovery thanks to better dispersion of the carbon nanotubes, and has potential applications for label control or proof marks in frozen foods.

III The **influence of NC on the mechanical and thermal properties of polyurethane** :

Polyurethanes have been used in a wide range [4] of applications such as the automotive, paint, furniture and textile industries. Although polymer composition varies from product to product, a covalent urethane bond is present to bridge the "hard" and "flexible" segments in a multi-block copolymer. The morphology of the two phases provides the key to controlling performance and multiple property tuning, by varying the composition or content of either phase.

Studies [4] show that the addition of small amounts (0.5-10% by volume) of multiwall carbon nanotubes (CNTs) to a thermoplastic elastomer, especially if the polyurethane[1] has a high hard segment content, produces nano-composite polymers with high electrical conductivity ($\sigma \sim 1'^{10}$ S / cm), low electrical percolation ($\varphi \sim 0.005$) and improved mechanical properties, including increased modulus and yield strength without losing elastomer strain at break above 1000%, and increased tensile strength.

Furthermore, due to intrinsic van der Waals forces [17], nanotubes are generally held together in bundles and have very low solubility in most solvents.

Functionalizing nanotubes leads to a slightly improved dispersion[1] for high percentages of carbon nanotube fillers, the improvements in this dispersion are small.

The average particle sizes of raw CNT carbon nanotubes [26] and carbon nanotubes treated with nitric acid A -CNT in aqueous solutions were 404.2 and 17.2 nm, respectively, indicating that the acid treatment led to a reduction in carbon nanotube agglomeration.

The properties of composites based on non-functionalized nanotubes were compared[1] to the properties of a film of composites based on SWNTC functionalized nanotubes bearing hydroxyl groups. This functionalization also destroys the intrinsic electrical conductivity of the nanotubes. However, composites manufactured and filled with higher fractions of functionalized nanotubes exhibit only slightly higher modulus in the rubbery plateau region. The addition of nanotubes slightly increases the glass transition temperature of the hard segment, but the glass transition temperature (Tg) of the polyurethane decreases[5] and the crystallization rate of the hard segments decreases with[1] a slight increase in the non-functionalized carbon nanotube content.

Filler dispersion (weight),[2] nanotube size ratio, and orientation are critical factors affecting the mechanical properties of particle-filled polymers. The interface plays an important role in composite reinforcement by effectively transferring stress between the CNT and the polyurethane matrix, even if the carbon nanotube size ratio has been reduced by surface modification.

For a system with high interfacial strength, the addition of carbon nanotubes (MWNT) to OH-terminated polyurethane increases the initial tensile strength of the composites.

Tensile tests[5] have suggested that multi-walled MWNT is more useful for improving modulus than single-walled carbon nanotube (SWNT), which is more favorable for improving polyurethane elongation.

Table 1 Hydrogen bonding index R and degree of phase separation (DPS) for PU–MWNT and PU–SWNT composites

CNT content	R		DPS	
	MWNT	SWNT	MWNT	SWNT
0%	0.86	0.86	0.46	0.46
0.5%	0.99	1.54	0.50	0.61
1%	1.08	1.55	0.52	0.61
2%	1.09	2.11	0.52	0.68

[5]

In general, the mechanical properties of fiber-reinforced composites are highly dependent on the degree of charge expansion (weight) between matrix and fibers. Thus, the thermal and mechanical properties of a PU-CNT composite are related to all the forces and interactions (Vander VALS, H-bonds, etc.), covalent bonds, ionic bonds, distribution, separation and size of the material's hard and flexible micro-phases, as well as its thermomechanical history and shape dimensions.

The literature contains dozens of reactions for functionalizing nanotubes, three methods for synthesizing PU-NTCs, and a dozen or so shaping techniques for obtaining an object article, which we have cited in this thesis.

1) *Influence of CNTs on the properties of composite films*

a) A study of the electrical conductivity[27] of an unfunctionalized PU- MWNTC-film formed as a melt and then extruded, shows that a much higher concentration of CNTs (3-4% by weight) is required to form an extruded composite-strand conductive network. This is explained in terms of the dynamic perlocation behavior of the CNTs network in the polymer melt.

b) Other PU-SWNTC-OH nano-composite films [26] (99.99 / 0.01 - 98.5 / 1.5) with improved thermal, mechanical and electrical properties were prepared, their initial tensile modulus, tensile strength and elongation at break of the PU-SWNTC film was 102 MPa, 6.6 MPa and 1.280%, respectively. The initial tensile modulus and tensile strength of the nano-composite film containing 1.5 wt.% PU-CNT were improved by around 19 and 12%, respectively, compared with the values corresponding to the original PU film. When CNT loading increased from 0 to 1.5% by weight, the tensile modulus and tensile strength of the PU-CNT nano-composite film increased from 102 to 121 MPa and from 6.6 to 7.4 MPa, respectively, however, elongation at break decreased from 1280 to-1243%.

The nano-composite film with PU **MWNTC-OH** containing 1.5 wt.% CNT had a conductivity of 1.2×10^4 S / cm, which was almost eight times higher than the order of magnitude of that of the pure PU film (2.5×10 to 12 S / cm). However, the conductivity values of PU-MWNTC nano-composites decreased exponentially with increasing CNT loading.

c) For a polyurethane-OH film with MWNTC-NH2 [2], when the carbon nanotube content reached 2.5% CNTs, tensile strength increased by 46%, and for 4% NC, tensile strength increased by 370%, and tensile modulus increased by 170.6%. Composites made with ester-functionalized nano-carbon (EST- SWNT) show a 104% increase in tensile strength over pure PU.

Thermal properties show [2] that the addition of carbon nanotubes would improve thermal degradation properties by 26^0 C ($315-341^0$ C) when the carbon nanotube content was 2.5% in a covalently bonded system.

d) For a film[3] with MWNT-CO-NH2, the glass transition temperature (Tg) of the polymer was greatly increased, from 5.4^0 C for the PU to 6.2^0 C for the composite. The improvement in Tg is mainly attributable to the excellent thermal properties of the CNTs and their direct bonds with the polyurethane hard segments. Above Tg, there is no obvious difference in the moduli for the two materials, while below Tg, the CNTs have a strong influence on the elastic properties of the PU matrix.

The temperature at maximum degradation rate increased from 408^0 C for PU to 419^0 C for the composite, showing that the thermal stability of the composite was improved by the addition of the carbon nanotubes, a fact that can be attributed to both the excellent thermal stability of the carbon nanotubes and their interactions with the polymer matrix.

At the breaking point on the stress-strain curves of the samples, the breaking strength of the composite is clearly superior to that of PU. The results showed that the carbon nanotubes in the polymer matrices play an essential role in load (weight) distribution. However, the elongation at break of the composite is slightly reduced, which may be caused by the uneven dispersion of CNTs in the medium.

e) For a film with MWNT-CO-NH-R-NH2[17] , experimental results showed that Tensile strength increased remarkably from 1.8 to 18 M.Pa, i.e. 900% higher than that of pure PU, following the addition of a MWNTs loading of 1% by weight of composites, then decreased with the addition of further loadings. Elongation at break increased by around 741% at 1% by weight MWNTs loading compared with the pure PU system. The results of the DSC measurements indicated that the Tg temperature of the flexible, PU- MWNT segment was altered, and new peaks near 54^0 C appeared, and they remained roughly constant as the MWNT content increased from 0.33 to 2.0% by weight. This result was observed due to microphase structure separation and molecular weight alteration of the PU hard and soft segments with the addition of MWNTCs.

2) _The influence of CNTs on the properties of composite yarns and fibers._

a) Polyurethane composites (PU- MWNTs) were prepared by electrospinning and solution molding. The morphological and thermal properties, as well as the mechanical performance, of nano-fibers and composite films were characterized and compared. Properties were compared between two types of material: a film and a nano-fibrous mat.[7].

The stress-strain curves for the film (N film or Cfilm) and nano-fibrous mat (Nmat or Cmat) for PU-MWNTC and pure PU show that Nmat had a tensile strength of 5.1 ± 0.2 MPa and an elongation of 540%. Its stress-strain curve was fairly monotonic in nature, showing a curve with a falling rate to failure. On the other hand, Nfilm showed a non-linear deformation behavior from 0% to 300% and a more linear curve at higher stresses. Nfilm had a tensile strength of 47.3 MPa, an elongation of 453% and a modulus of 18.3 MPa. Comparing the two, Nfilm had a 9-fold higher tensile strength than Nmat. In addition, Nfilm had a much higher modulus and elongation at break. The difference in their mechanical performance could be explained by their morphologies; the low density of the nanofibrous mat, with its many pores, could have affected its mechanical properties. During the tensile test, the fibers that were oriented along the main deformation axis, for the Nfilm obtained, it is solid and throughout, mechanically isotropic, therefore, it showed a characteristic curve of an elastomeric material.

When MWNTs were incorporated into the PU matrix, (Cmat) showed an increase in tensile strength, from 5.1 to 8.6 MPa, an increase of 69%. Modulus also showed a more than 2-fold increase from 1.5 to 3.6 MPa, but flexibility decreased slightly from 540% to 453%. At the same time, Cfilm showed a similar curve pattern to Nfilm, but had superior mechanical properties. Tensile strength, elongation and modulus increased by 62%, 24% and 78%, respectively, compared with Nfilm.

For effective load (weight) distribution from the PU matrix to the MWNTs, there should be interfacial stress between the MWNTs and the PU. MWNT orientation and aspect ratio are also important factors in this case. In general, MWNTs in straight form, and homogeneously dispersed in the host polymer, will provide better load distribution than MWNTs in looped and agglomerated form. Incorporation of MWNTs reduces transparency for films and nanofibers - MWNTs/PU showed improved thermal degradation behavior, with the incorporation of a low MWNT content (0.1%).

In this study[7] , the MWNTs were homogeneously dispersed in the PU film, showing individual MWNTs enveloped by the PU matrix, but they were in curly shapes. On the other hand, MWNTs in nano-fibers are generally aligned along the axis of the nano-fiber, particularly along its surface, but there was a non-uniform distribution of MWNTs in the nano-fiber. This could explain the lower elongation of CmaT compared with nMat. In addition, there could be some agglomeration of MWNTs in the

nano-fibers, which could act as stress concentration points leading to failure. The increase in tensile strength for both; the PU- MWNT film and the nano-fibrous mat could be due to the following:

(A) After acid treatment of MWNTs, some form of defect was observed on the MWNT surface. These defects could act as anchoring sites for PU locking.

(B) acid treatment would graft certain carboxylic acid groups, which can improve the interaction of MWNTs with the OH groups of the PU chain.

The thermal stability of PU/MWNT composite nano-fibrous films and mats showed an improvement over their pure Pus counterparts. It should be noted that the MWNT content present was only 0.1.%.

A much better improvement in the thermal stability of composites could be possible by incorporating higher quantities of MWNTs. This increase could be attributed to the excellent thermal stability of MWNTs, which were fully enveloped by the PU matrix. Acid treatment of MWNTs, which attaches functional groups to the MWNT surface, also contributed to the good interaction between MWNTs and PU as verified by Raman and FTIR results. In addition, differential scanning calorimetry (DSC) measurements show some shifting of the endothermic peaks of PU/MWNT composite films and nano-fibrous composite mats, especially at 170-180⁰ C, which is associated with melting of the PU hard segment. This suggests some interactions of the hard segment with the MWCNT nanotubes, by means of a newly formed hydrogen bond, which stabilizes the dynamic thermal properties of the composite films.

b) A thermoplastic polyurethane (TPU) elastomer was used [9] to manufacture composite fibers with MWNT-COOH, and increasing the nanotube concentration from 0 to 17.7 wt.% resulted in a monotonic increase with a maximum at around 9.3 wt.%. At this concentration, the tensile strength was around 2.4 times greater than that of pure PU fiber. The decrease in strength at high CNT concentrations such as 17.7 wt.% can be attributed to the increased frequency of localized aggregations, which can be revealed by scanning electron microscopy (SEM) of composites at higher concentration levels.

A significant improvement in Young's modulus and tensile strength was achieved by incorporating MWNTs up to 9.3% by weight while not sacrificing a high elongation of Polyurethane elastomer at break[9] . The results showed a homogeneous dispersion of MWNTs throughout the polymer matrix and strong interfacial adhesion between oxidized MWNTs and the matrix, resulting in[9] the considerable improvement in the mechanical properties of composite fibers.

c) The composite was prepared using a PU filter membrane and carbon nanotube[25] by electrospinning, and can be stretched by up to 400%, during which the electrical resistance is increased by 270 times. The composite is sensitive to compression and to the vapor of organic solvents, and is used as a highly deformable layer for chemical vapor detection, flexible electromagnetic shielding and lightning protection.

3) *Modification of the properties of PU-NTC composites into foam shapes.*

a) MWNT-reinforced PU thermoplastics have also been[12] prepared by the melt-blending method followed by compression molding. Important factors for the production of carbon nanotube-reinforced polymers are also widely established, such as the homogeneous dispersion efficiency of carbon nanotubes in a polymer matrix and the introduction of a strong interfacial interaction between the polymer matrix and the CNTs, to form an effective stress distribution characteristic of interphases.

The results show that the interfacial interaction between the TPU matrix and the MWNTs increases with increasing %wt CN. Hydrogen bonding between the C $=$O groups of the urethane bonded to the TPU hard segments and a hydrogen atom of the -COOH groups of MWNT is well favored. The existence of the interfacial interaction between TPU and MWCNT significantly enhanced the properties of TPU / MWNT nano-composite materials.

The TGA curves for TPU / MWNTs nano-composites shift towards a higher temperature compared to the pure polyurethane matrix, indicating that CNT incorporation increases the peak temperature of the maximum degradation curve velocity. Thermal degradation of TPU takes place in two stages, due to the presence of two segments within the matrix. The first and second stages of thermal decomposition correspond to the hard and flexible segments respectively. The degradation temperature peaks, at maximum speeds, in the DTG curves, for hard and flexible domains of the pure polyurethane matrix, are around 321 and 408^0 C, they were improved to around 347 and 418^0 C, respectively, when the TPU nano-composites will be loaded with 2.5% by weight of MWNT-COOH.

Thermo-oxidative degradation of both hard and flexible TPU matrix segments is successfully retarded by the addition of MWNT, indicating that CNT interacts uniformly with hard and flexible segments within the TPU molecular structure without any preferential domain association.

The glass transition temperature (Tg) of the flexible segment phase (Tg (flexible)) of pure TPU is around -71 °C. The Tg values of TPU nano-composites filled with 0.5, 2.5 and 5.0% by weight MWNT are -73, -72 and -73 °C, respectively. The Tg of MWNT-filled TPU nano-composites is slightly lower than the Tg of pure TPU.

The tensile strength and modulus of rupture of TPU nano-composites filled with 0.5% MWNTs, incremented to an optimum level, and a further addition of CNT obviously reduces tensile strength.

The initial modulus, tensile strength and elongation at break of TPU- 0.5% by weight MWNTs nano-composites were 3.43 MPa, 16.00 MPa and 1392, respectively, and which are improved by around 15.10%, 64.27% and 13.37%, respectively compared to the pure homologous matrix.

The elongation at break of TPU- MWNTs nano-composites increases compared to that of the pure TPU matrix, up to a CNT loading of 2.5% by weight, due to the dispersion of CNT tubes in the TPU matrix. But it decreases when the MWNT content is 5.0%, due to the poor dispersion of CNTs in the TPU matrix.

The shear strength of the pure TPU matrix is 40.56 N / mm, while that of the TPU matrices loaded with 0.5, 2.5 and 5.0% MWNT by weight are 60.00, 69.30 and 59.60 N / mm, respectively. It can be seen that the shear strength of TPU nano-composites increases significantly with increasing MWNT loading, which may be due to the high torsional energy of carbon nanotubes in the TPU matrix.

The shear strength of nano-composite TPU- 2.5% by weight MWNT is 69.30 N / mm, which is about 70.86% higher than that of clean TPU.

The increase in hardness is probably due to the reinforcing effect of uniformly dispersed MWNTs, the shear strength significantly increased up to a TPU- MWNTs nanocomposite loading of 2.5% by weight, but it is considerably deteriorated at 5.0% MWNT loading by weight, due to the aggregation of some of the CNTs in the TPU matrix.

Material hardness indicates a material's inherent ability to limit deformation - localized deformation caused by the application of external stimulation. Hardness is directly associated with the interconnected and cross-linked networks, elasticity, plasticity, strength, modulus and porosity of the polymer matrix.

The hardness of TPU- 5.0 wt% MWNTs nano-composites improved by around 15.28% compared with the pure TPU matrix. The hardness of TPU- MWNTs nano-composites increased with increasing MWNT.

Rheological analysis showed the low shear modulus and shear-thinning behavior of nano-composite.

b) The preparation of CNT/ PU rigid[23] (composite foam) with a low perlocation threshold for 1.2% by weight CNT is achieved by constructing efficient conductive tracks with a homogeneous dispersion of CNTs. The conductive foam exhibits excellent electrical stability under various temperature fields. In the presence of 2.0 wt.% carbon nanotubes, the cell size shows a slight increase, indicating that CNTs cannot serve as well as a heterogeneous nucleating agent during the foaming process. Highlighting potential applications for long-term use over a wide temperature range from 20 to 180°C. Compression measurement and dynamic mechanical analysis indicated a 31% improvement in compression properties and a 50% increase in storage modulus at room temperature in the presence of 2% CNT. The incorporation of only 0.5% by weight of CNT induced remarkable thermal stabilization, and the degradation temperature increased from 450 to 499°C at 50% weight loss [23].

4) *Influences of CNTs on the properties of SMP shape memory composites*

a) Shape memory polymers (SMPS)[6] have been one of the most studied subjects in recent years, because they can be deformed and fixed in a temporary shape. They can also recover their permanent shape by responding to various external stimuli - heat, light, electricity, magnetic field or dissolution, for example. Polyurethane block copolymer (PU) is one of the most widely used materials in this field, because its chemical structure can be easily controlled, and because of its ability to recover its shape by changing temperature. PU is a kind of elastic thermoplastic made up of flexible and hard segments. These elastic materials appear in two distinct phases. Its shape memory effect is generally induced by thermal stimulation.

By heating PU above the transition temperature (the glass transition temperature or melting temperature), the flexible segments can be used as molecular switches and are responsible for shaping the recovery of temporary deformed shapes.

The permanent shape memory[14] of the hard segment can be physical or chemical in nature, such as crystals, glassy domains, chain entanglements, or chemical cross-links. The flexible segment works as a reversible molecular switch that controls the cycle of the shape memory process.

The switching temperature of the shape memory effect ranges from 44 to 55^0 C, depending on the content of flexible segments, which varies between 50 and

90% by weight. The molecular weight of the flexible segment must be low when low temperatures are applied.

The electroactive shape memory effect of PU-SWCNT hybrids was induced by the release of thermal energy. The PU-SWCNT hybrid acts like an electrical resistor. When voltage was applied, current passed through the resistor, consuming the power and thermal energy produced. The temperature of the resistor surface then rises.

The electroactive shape memory effect became more active when the induced thermal energy exceeded the flexible segment melting point of the PU-SWCNT hybrids.

The optimum conduction condition for PU-SWCNT hybrid material networks was found when 3% SWCNTs were incorporated into the PU. PU-SWCNT hybrids containing 4% SWCNTs showed good electroactive shape recovery properties when an electrical voltage was applied at low temperature. Recovery of the original shape of the sample was around 88% for 90 s when an electric current of 30 V and 0.1 Amp was triggered.

Further analyses[6] have shown that SWCNT has the effect of varying the melting temperature detected by DSC, and influences the crystallization of PU-SWCNT composites.

b) Incorporating nanotubes into the matrix of a Hyperbranched Polyurethane (HBPU)[10] significantly improves mechanical properties due to the well-known nano-strengthening effect of CNTs. Pure HBPU shows a tensile strength of 6 MPa, while the composite comprising 1.% of its weight of MWCNT has a strength of 25 MPa. With an increase in the MWCNT dose from 1 to 5.% by weight, the strength also rises to 46 MPa. In general, HBPU lacks mechanical strength due to their low viscosity and low chain entanglement rate. An approximately 3-fold increase in tensile strength was achieved by the formation of composites, even at very low MWCNT loading (1.wt.%).

Inserting these MWCNTs into the hyperbranched PU matrix increases its thermal stability. Pure HBPU was stable up to 215^0 C, while incorporating 1.wt.% MWCNTs into the matrix would increase thermal stability up to 275^0 C. In addition, the composites showed a single-step degradation, whereas pure HBPU exhibited a two-step degradation profile. The reason for the steady improvement in these parameters is due to the fact that MWCNTs act as physical cross-linking points, which limit the movement of polymer chains.

MWCNTs also act as thermal insulators, and the delay in thermal degradation is mainly due to the presence of thermally insulating MWCNTs well dispersed in the polymer matrix. Enhanced interfacial adhesion between modified MWCNTs and HBPU would delay segmental movement of polymer chains. The increase in residual yield with increasing MWCNT loading between 1 and 5.wt.%, further demonstrates the heat-insulating nature of the nanofiller.

The melting temperature of the soft segments for composites was increased from 47 to 56^0 C compared with that of the pure polymer. The dispersion of the nanotubes, as well as the formation of the rigid structure via different molecular interactions is responsible for this increase. Increasing the amount of MWCNT also raises the melting temperature.

Crystallinity increases with the incorporation of MWCNTs as a result of an increase in the number of unblocked oriented chains. The same factors that influence crystallinity also influence shape memory properties. A shape recovery of 84% is observed for HBPU / MWCNT composites with a loading of 2.wt.% MWCNT, while this value is 97.3% for HBPU/MWCNT of 2.5% in the present study[10] .

c) Films of electroactive shape-memory PU-MWNTs composites were synthesized [13] from PU containing 40% hard segments. Modulus and tensile strength at 100% elongation increased with increasing content of modified MWNTs, while elongation at break decreased.

Modification of the MWNT surface also resulted in a decrease in the electrical conductivity of the composite materials, however, when the MWNT content in the matrix increases, the conductivity increases and an order of $10^{.3}$ S.cm^{-1} was obtained for samples with 5% by weight of modified MWNT.

d) One series, PU-MWCNTOH-PCL [14] with different MWCNT weight contents (2.4, 3, 4, 5, 10, 12%) were synthesized by ring-opening polymerization of a capro-lactone, the MWCNTs surface was linked by chemical bonds to PCL chains. SEM images, show a homogeneous dispersion of carbon nanotubes in the PU matrix, even with increased MWCNT loading, the nanotubes could still be evenly distributed in the polymer matrix and no aggregation is observed. The material's tensile strength of over 240 MPa and elongation at break ranging from 750 to

1400% both showed extraordinarily high values. MWCNT with homogeneous dispersion in the PU matrix achieves high mechanical properties, resulting in recovery ratios (Rr) of 97% and cracking ratios (Rf) of 94%.

e) Shape memory polyurethanes (SMP)[16] exhibit a shape memory effect resulting from the thermodynamic incompatibility between hard segments (aromatic di-isocyanates) and flexible segments (polyethers or aliphatic polyesters)... The influence of MWNTs insertion on the shape recovery effect is remarkable. The shape recovery rate of composites loaded with 1.0 and 3.0 wt.% MWNTs is higher than that of pure SMP fiber. However, at a loading of 7.0% by weight, the shape recovery ratio is much lower than that of pure SMP. Maximum stress at 100% strain increases monotonically. The recovery ratio increases from 83% (pure SMP fiber) to 91% (fiber with 1.0% by weight MWNT).

The corresponding SMP fibers were prepared by melt-spinning[16] . The spinning capacity of SMP- MWNTs decreased significantly with increasing MWNT content. When the MWNT content reached 8.0% by weight, the fibers could not be produced due to the poor rheological properties of the composite materials. Crystallization in SMF was favored at low MWNTs because it acted as a nucleating agent. However, at high MWNT contents, crystallization was prevented due to the limited movement of the polyurethane chains. Homogeneously distributed and aligned MWNTs preserve the SMF's high initial toughness and modulus. Recovery rate and recovery strength were also improved because MWNTs helped store internal elastic energy during stretching and shaping.

f) For a new series [21] of electroactive shape-memory polyurethane composites, a twofold improvement in the thermomechanical properties of samples prepared by in situ polymerization was observed; elongation of up to 200% compared with that of conventional samples. The resulting composites, with 92% and 95% shape recovery, could be used as preferred materials in various actuators. **An ester-based polyurethane (PU)** [28] **with a low glass transition**

temperature was used, and oxidized carbon nanotubes (ox - CNTS) were introduced by melt blending. The polymer matrix used in this work is a kind of linear ester-based polyurethane that forms hydrogen bonds with urethane, the phases are connected by hydrogen and urethane bonds and probably by hydrogen bond bridges between the ester groups and the -OH functions of the flexible polyol segments, and with the -COOH functions of the oxidized carbon nanotubes.

The results show that improved dispersion of oxidized CNTs contributes to high stiffness below the glass transition temperature (Tg), while storage modulus (E_0) decreases above T_g. However, the first stress at 50% strain for three samples, PU-CNT, PU and PU-ox-CNT, were 4.04 MPa, 5.24 MPa and 5.78 MPa, respectively. The incorporation of pristine CNTs resulted in a slight decrease in modulus of elasticity, while tensile strength and elongation at break increased due to the incorporation of CNTs and ox-CNTs. The nano-composites show a shape recovery rate in excess of 98% [28].

Chapter 2. NEW POLYURETHANE-CARBON NANOTUBE COMPOSITES: SYNTHESIS, REACTIONS AND PROCESSING

A. Preliminary tests

Obtaining a PU material with homogeneous texture and consistent mechanical and thermal behavior depends on how carbon nanotubes are incorporated into polyurethane networks. In a number of preliminary tests, we added quantities of polyol, nanotubes, isocyanates and D.B.T.L. by simple physical mixing; and for different proportions of nanotubes, PU samples shaped in this way did not show monotonic thermal stability profiles, although the carbon nanotube loadings did not exceed a percentage of 12%.

In a second series of tests, we prepared a masterbatch with a high carbon nanotube concentration of 24% of the polyol mass and subjected it to cold stirring for 40 minutes. The other polyols, which are loaded with carbon nanotubes at lower percentages, will be made from this polyol masterbatch, by taking samples of different volumes, added to calculated quantities of pure polyol. Each sample is manually stirred and homogenized for 10 minutes. The PU samples obtained in this case do not dry out even after two days of their molding, they are not very viscous and sticky, and the carbon nanotubes crumble from the material, especially for PU containing a carbon percentage greater than 12%. An increase in di-isocyanate mass of 12.5% will be necessary to obtain dry, stable PU.

B. Polyurethane preparations containing non-functionalized carbon nanotubes

Taking this preparatory work into account, we succeeded in developing six chemically different families of polyurethanes, and each family comprises a number of similar samples differing only in their mass content of carbon nanotubes.

1) **1ère family**: Samples in this family are prepared from the **type I**

masterbatch (see chapter 3, page...), including carbon nanotube-filled polyols in lower percentages. Samples of different volumes, added to calculated quantities of pure polyol, and each manually stirred for 10 minutes, are added to the di-isocyanates and catalyst, depending on the quantities used, as shown in Table (1). Isocyanates are characterized by the - N=C=O function, which is highly reactive due to its two unsaturations. This reactivity is due to the electrophilic nature of the carbon atom, which gives the isocyanate group a structure that allows for resonance possibilities. The reactions that would take place are given in the following equations (fig-16)

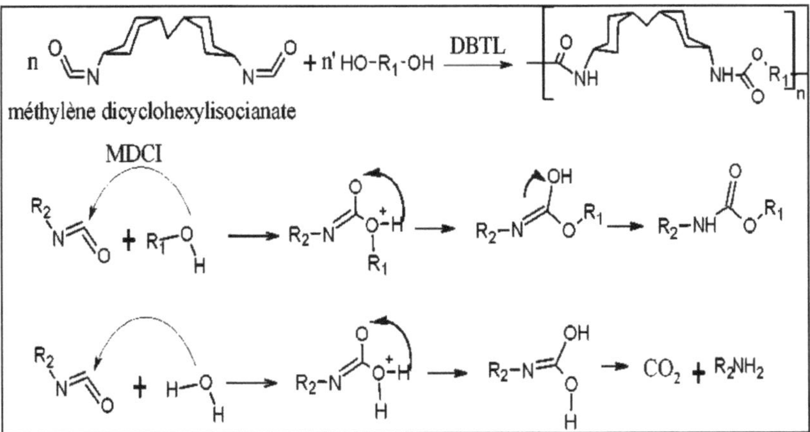

Figure 21- Chemical reaction equations for polyurethane processing

After stirring the reaction mixture by hand for 5 minutes, a fraction of the mixture is sampled in each case. evolution (approx. 10ml) for use in preparing samples for the second family. The large remainder will be poured into a vase to react and swell at rest.

Table 1- Sample preparation HSPU-1750- Carbon nanotubes

PU names	Mass of mother solution in gr	mass of Polyol RO-R	NCO mass in gr		Water	D.B.T.L.	Carbon nanotube mass	C/PU%	% in C/OH
HS-PU1750-24%	29.76	0.0	18gr		+	+	5.76	12.6 %	24%
HS-PU1750-16%	19.84	8.0	18gr		+	+	3.84	8.8%	16%
HS-PU1750-12%	14.88	12gr	18gr		+	+	2.88	6.7%	12%
HS-PU1750-3%	3.72	21gr	18gr		+	+	0.72	1.8%	3%
HS-PU1750- 0.0% HS-PU1750- 0.0% HS-PU1750- 0.0	0.0	24	18gr	Mechanical shaking ;10 mn	+	+	0.0	0.0%	0.0%

2) 2ème family: PUs in this family are derived from the first; 3 drops (approx. 2ml) of trimethoxyphenyl silane, 2g of isocyanate and 2 drops of D.B.T.L are added to the fraction of the evolving mixture already removed, and stirred for 5 minutes before molding. Table (2) shows the procedure for preparing PU containing Si.

Table 2-Preparation of HSPU1750-NCO-Si samples

sampling	Trimethoxy phenylsilane	NCO	manual gitation; 5 mn	D.B.T.L.	C/OH	Samples from the 2ème family
HS-PU1750-24%/ 10ml	3drops	2gr		+	24%	HS-PU1750-Si-24%
HS-PU1750-16% /10ml	3drops	2gr		+	16%	HS-PU1750- Si- 16% - French
HS-PU1750-12% /10ml	3drops	2gr		+	12%	HS-PU1750- Si- 12% - Si- 12% - Si- 12% - Si- 12

HS-PU1750-3% /10ml	3drops	2gr	+	3%	HS-PU1750- Si-3% - Si-3% - Si-3% - Si-3% - Si-3% - Si-3% - Si-3
HS-PU1750- 0.0% HS-PU1750- 0.0% HS-PU1750-0.0 /10ml	3drops	2gr	+	0.0 %	HS-PU1750- Si- 0.0

Samples derived from them are designated by a prefix [HS -PU-1750-...-...] for the first family and [HS -PU-1750-...-si] for the second.

The addition of trimethoxyphenyl silane can cause NCO end coupling reactions, which may influence the rate of bridging bonds, and consequently the rigidity of the resulting material and its thermal stability. The synthesized samples will be hardness tested and examined by TGA.

Material hardness indicates a material's inherent ability to limit deformation - localized deformation caused by the application of external stimulation. Hardness is directly associated with the interconnected and cross-linked networks, elasticity, plasticity, strength, modulus and porosity of the polymer matrix.

C. Preparation of polyether polyols containing functionalized carbon nanotubes by oxidation followed by esterification

The dispersion of nanotubes in the PU matrix was always a key issue. Functionalization of nanotubes leads to better dispersion at high carbon contents. [1] Buffa, F., & Resasco, D. (2007) used functionalized nanotubes containing hydroxyl groups to improve interaction with urethane groups via hydrogen bonding. The study of the composite structure shows that carbon nanotubes can be dispersed in the polyurethane matrix... Another group of researchers [3] Xiong, J.& Wang, X. (2006) succeeded in attaching a COOH carboxyl group to the nanotube by oxidative treatment, and the chemical bonding of carbon nanotubes to the polyurethane matrix was confirmed by

Fourier transform infrared spectra.

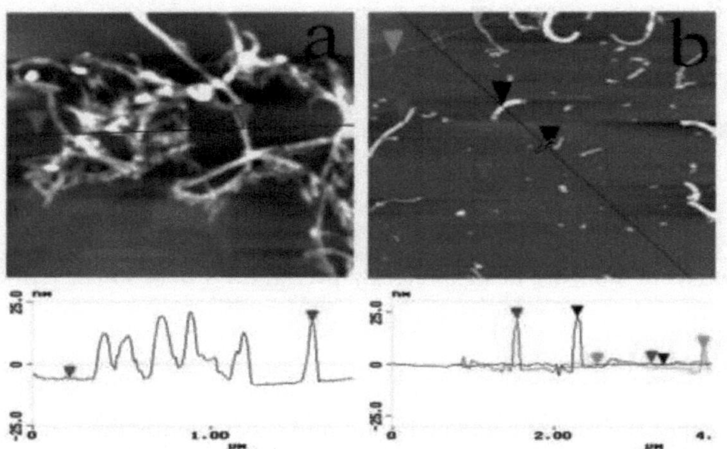

Fig. 1. AFM images of CNTs: (a) as-received carbon nantubes and (b) acid-treated carbon nanotubes.

Figure 22- Images of carbon nanotubes before (a) and after oxidizing acid treatment 1)Family 3 and family 4 are made from two **type II** parent mixtures, characterized by functionalization of carbon nanotubes with a sulfo-nitric mixture followed by esterification with citric acid hydroxyl.

Two **type II** masterbatches containing 24% by weight of functionalized nanotubes are prepared, one with polyether polyol (M=1200gr, f=3) and the other with polyether polyol (M=1750gr, f=3). (See Chapter 3). The resulting samples are designated with a prefix [AN -PU-1200-....-...-...] and [AN -PU-1750-....-...-...].

2) Two further mother mixtures **type III** are made from the oxidized nanotubes and esterified with hydroquinone and the two polyols will be invested for the preparation of two families 5 and families 6. The resulting samples are designated with a prefix [KC -PU-1200....-...-...] and [KC -PU-1750........-...].

The four polyols obtained and used for PU synthesis are :

*   Carbon nanotubes 24%,($H+$)/Citric acid, Polyol, (M=1200, f=3).

*   Carbon nanotubes24%,(H^+) /Citric acid Polyol, (M=1750, f=3).

- Carbon nanotubes24%,(H⁺) /Hydroquinone Polyol, (M=1200, f=3).

- Carbon nanotubes24%,(H⁺) / Hydroquinone Polyol, (M=1750, f=3).

3) Probable chemical reactions :

In the sulfonitric mixture, the number of moles of pure sulfuric acid (n=0.179mole) is in excess of the hydroxyl of citric acid (0.04487 mole - 4 times) and again in excess of the hydroxyl of citric acid (0.04487 mole - 4 times).

Hydroquinone diol (2x0.022= 0.044 mol-4 times).esterification reactions are favored at temperatures above100 C.⁰

a) Sulfonitrile treatment of carbon nanotubes

$$\square \quad + \quad H_2SO_4/HNO_3 \quad \xrightarrow{\text{partielle}} \quad \text{(carboxylic functionalized nanotube)}$$

b) Esterification with citric acid hydroxyl

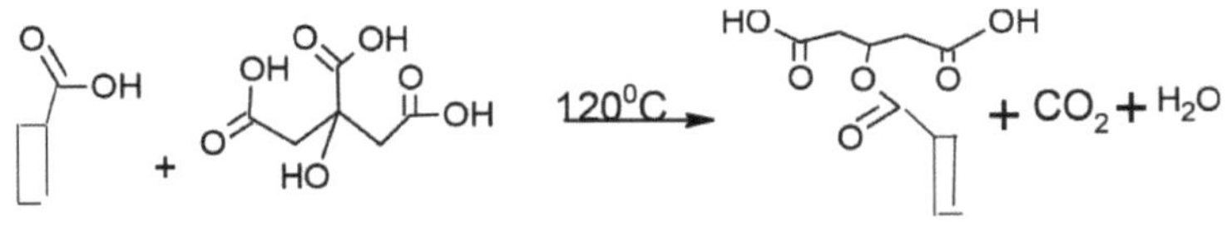

$$\text{(carboxyl nanotube)} + \text{(citric acid)} \quad \xrightarrow{120^0C} \quad \text{(product)} \quad + CO_2 + H_2O$$

Both types of nanotube are functionalized either by carboxylic acid functions for treatment with citric acid, or by alcoholic OH functions.

The reaction of isocyanates with carboxylic acids leads to the formation of amides, again with the release of co2, but the electrophilic nature of the COOH function considerably limits its rate, compared with that of alcohols.

Jung et al [8] crosslinked polyurethane (PU) chains to MWCNT nanotubes by a reaction between carboxylic acid groups of oxidized MWCNT and isocyanate (NCO) groups of prepolyurethane.

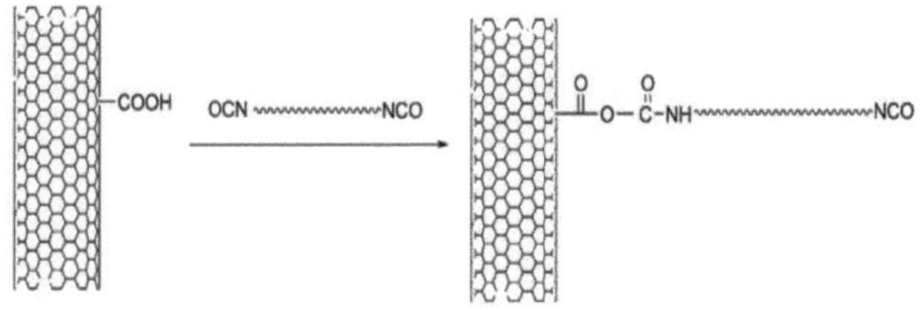

Fig. 6. Crosslinking of polyurethanes chains with carbon nanotubes.

[24]

Figure 23-Linkage between polyurethanes and carbon nanotubes

4) The preparation of polyols loaded with different percentages of carbon nanotubes is carried out by diluting one of the stock solutions, depending on the family; the polyols, loaded with carbon nanotubes with percentages 18, 12, and 6% by mass, will be made from each polyether polyol stock mixture, by taking samples of different volumes, adding them to calculated quantities of pure polyol, and well homogenized for 24 hours by mechanical agitation of a shaker rotating at a speed of 500 rpm.

D. <u>Polyurethane preparation</u>

The quantities of polyols indicated in tables 4, 5, 6 and 7 are added, each time, to 25 g of isocyanates, 1 ml of water and 3 drops of tin dibuthyl laureate (D.B.T.L.).

Table 3- Preparation of PU with oxidized nanotubes esterified with citric acid/polyether polyols (M=1200 f=3)

EXPERIENCE	Mass of type II mother mix in gr	Mass of pure polyol added	Cyclic isocyanate mass in gr	Mass of aliphatic isocyanate in gr	H2O volume in ml	DBTL drops	Comments
AN-PU1200-25-NCO(*)- 24%-A	25	0	25	0	1	3	viscous
AN-PU1200-20-NCO(*)- 24%-A	20	0	25	0	1	3	putty
AN-PU1200-18- NCO(*)- 24%-A	18	0	25	0	1	3	flexible

AN-PU1200-18- NCO(*)- 18%-A	13.5	4.5	25	0	1	3	flexible
AN-PU1200-18- NCO(*)- 12%-A	9	9	25	0	1	3	flexible
AN-PU1200-18- NCO(*)-6%- A	4.5	13.5	25	0	1	3	flexible
AN-PU1200-18- NCO(al)- 24%-A	18	0	0	25	1	3	rigid
AN-PU1200-18- NCO(al)- 18%-A	13.5	4.5	0g	25	1	3	rigid

The stoichiometry or ratio $\frac{NCO}{OH\ total} > 1 \cong 1.2,$ as well as the structure of the carbon skeleton carrying the NCO functional groups, determine the physical appearance of the polyurethanes obtained; the addition of excess quantities of polyols (20 or 25gr) resulted in viscous, sticky and soft polyurethanes, with dicyclohexyl methane 4,4'-di-isocyanate H12MDI, while samples prepared with 18gr of polyols are soft and flexible.

Other samples prepared using the same ratio $\frac{NCO}{ROH} = \frac{25gr}{18gr}$, but with aliphatic di-isocyanates are rigid.

The necessary increase in the ratio $\frac{NCO}{ROH}$. is made to compensate for the consumption of isocyanates by reactions linked to the presence, in the reaction medium, of either OH groups originating from citric acid, and which are not completely esterified, or traces of water originating from the humidity of the nanotubes.

Table 4- Preparation of PU with nanot citric ubes oxidized and esterified by acid/polyether polyols (M= L750 f=3)

	Mass of type II mother mix in gr	Mass of pure polyol added (1750) in gr	in gr	H20 volume in ml	DBTL drops	Comments
AN-PU1750-18- NCO(*)-24%- A	18	0	25	1	3	flexible
AN-PU1750-18- NCO(*)-18%- A	13.5	4.5	25	1	3	flexible
AN-PU1750-18- NCO(*)-12%- A	9	9	25	1	3	flexible
AN-PU1750-18- NCO(*)-6%-A	4.5	13.5	25	1	3	flexible

Tables {4, and 5} show the synthesis methods for various polyurethanes whose polyols are treated with citric acid:

1) A further increase in the ratio $\dfrac{NCO}{OH} = \dfrac{25gr}{16gr}$ is required to prepare polyurethanes whose polyols are treated with hydroquinone; The esterification reaction is excluded can give; These phenolic ends remaining from the primary hydroquinone, will react with isocyanates.

Tables {5, and 6} show the synthesis methods for various polyurethanes whose polyols are treated with hydroquinone:

Table 5- Preparation of PU with nanotubes oxidized in the presence of hydroquinone / polyether polyols (M=1200 f=3)

EXPERIENCE	4 mass of type II mother mix in gr	"lasse de polyol pur jouté (1200) en gr	"cyclic isocyanate mass ;n gr	"iliphatic isocyanate mass in gr	H20 volume in ml	)BTL in drops	)bservations

KC-PU1200-25-NCO (*)- 24%-A	25	0	25	0	1	3	viscous
KC-PU1200-20-NCO (*)- 24%-A	20	0	25	0	1	3	putty
KC-PU1200-16-NCO (*)- 24%-A	16	0	25	0	1	3	flexible
KC-PU1200-16-NCO (*)- 18%-A	12	4	25	0	1	3	flexible
KC-PU1200-16-NCO (*)- 12%-A	8	8	25	0	1	3	flexible
KC-PU1200-16-NCO (*)- 06%-A	4	12	25	0	1	3	flexible
KC-PU1200-16-NCO (al)- 24%-A	16	0	0	25	1	3	rigid
KC-PU1200-16-NCO (al)- 18%-A	12	4	0g	25	1	3	Rigid

Table 6- Preparation of PU with nanotubes oxidized and esterified with hydroquinone and polyether polyols (M=1750 f=3)

EXPERIENCE	Mass of type III mother mix in gr	Mass of pure polyol added (1750) in gr	Cyclic isocyanate mass in gr	volume H2O	DBTL gear	Comments
KC-PU1750-16- NCO (Φ)- 24%-A	16	0	25	1	3	flexible
KC-PU1750-16- NCO (Φ)- 18%-A	12	4	25	1	3	flexible
KC-PU1750-16- NCO (Φ)- 12%-A	8	8	25	1	3	flexible
KC-PU1750-16- NCO (Φ)- 06%-A	4	12	25	1	3	flexible

Chapter 3. NEW POLYURETHANE-CARBON NANOTUBE COMPOSITES

1) *Homogeneity study by scanning electron microscopy*

The homogeneous dispersion of CNTs in the PU matrix is a necessity on several levels, especially for: thermal stability, tendency to crystallize, glass transition temperature, good reinforcement of the material's mechanical strength, especially tensile strength, and elongation at fracture. Heterogeneities can lead to structural defects in the composite material. Figures (25, 26, 27, 28, 29) show images, taken by MBE scanning electron microscopy, of the fracture surfaces in cross-sections of the prepared PU-NTCs composite materials. The bright spots and lines are the ends of the broken or aggregated CNTs. The size and shape of the light spots are related to the orientations of the nanotube axes with respect to the fracture surface plane.

The figures show that the distribution of CNTs in composites from two families, HSPU-1750-NTC and HSPU-1750-NTC-Si, was not uniform.

On the other hand, the two composites **AN-PU1200-18- NCO(*)-24% (FIG-27)** and **AN-PU1750-18- NCO(*)-24% (FIG-29)** show homogeneous textures and few aggregated particles. Equally homogeneous textures are expected for the other composites derived by simply diluting the {polyol-24%NTC} blends to obtain polyol-NTC blends containing 18%, 12% and 6% nanotubes by mass, Figures **(FIG-29)** show very satisfactory dispersion in the morphology of the two foams {AN-PU1750-18 **NCO(F)-6%} and {AN-PU1750-18NCO(F)-12%},** less satisfactory dispersion for the {AN-PU1750-18NCO**(F)-18%}** composite. Aggregations are noted in the micrographs of **{AN-PU1200-18NCO(F)-18%}** **and a** relatively well-dispersed **texture** in the matrix of {AN-PU1200-18NCO**(F)-**

12%} samples **(FIG-27).**

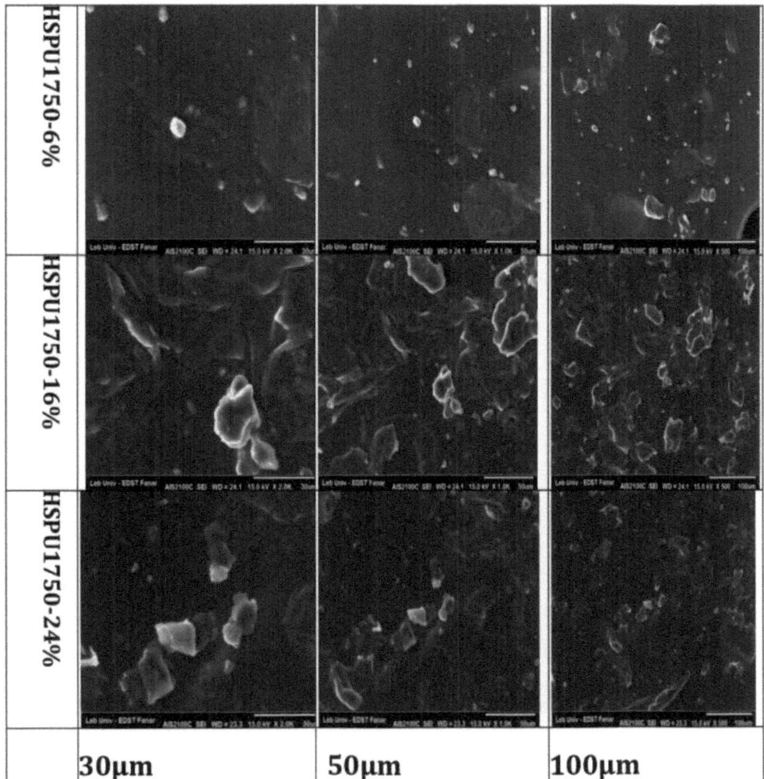

Figure 24- MBE micrographs of compositesHS 1750- 24-NCO18 (Scanning Electron Microscopy =SEM kV : 30.00 Tilt: 0.00 Take-off:23.42 ; AmpT : 12.8 Detector Type: SDD Apollo X ; Resolution :130.29 Lsec 52 EDAX ZAF Quantification (Standardless); Element Normalized).

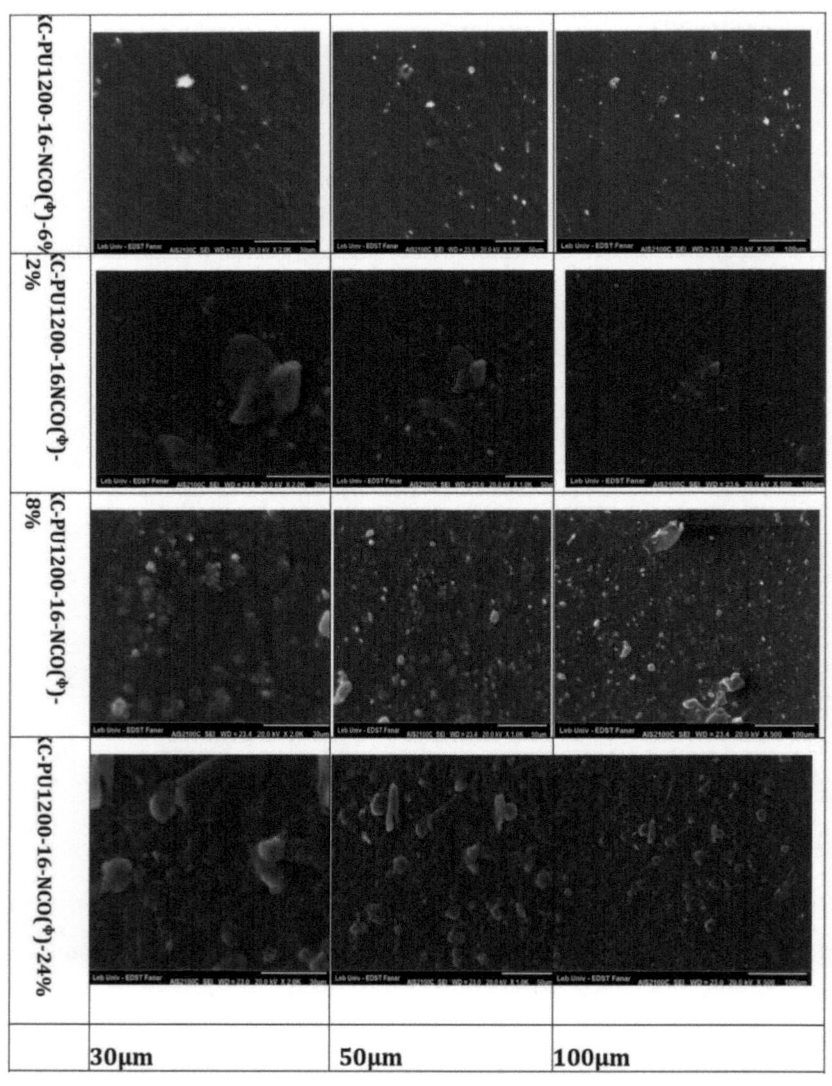

Figure 25-MBE micrographs ofKC1200-16-NCO composites

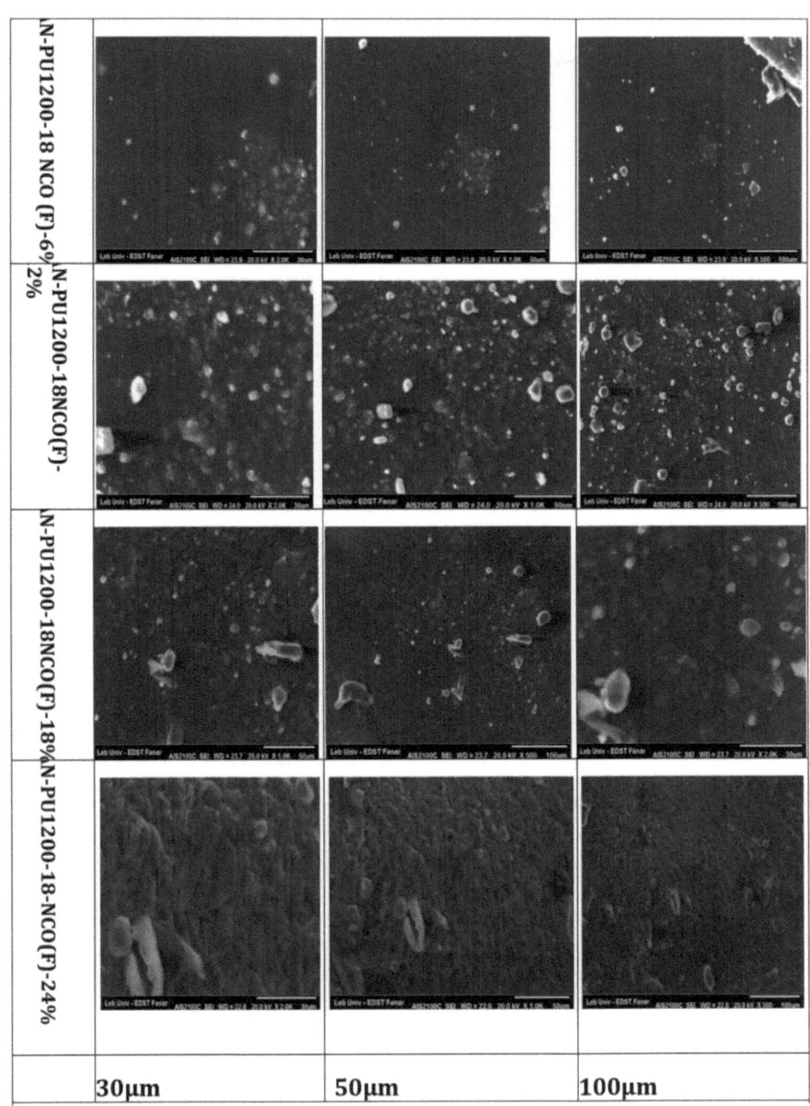

Figure 26-MBE micrographs ofAN1200-18-NCO composites

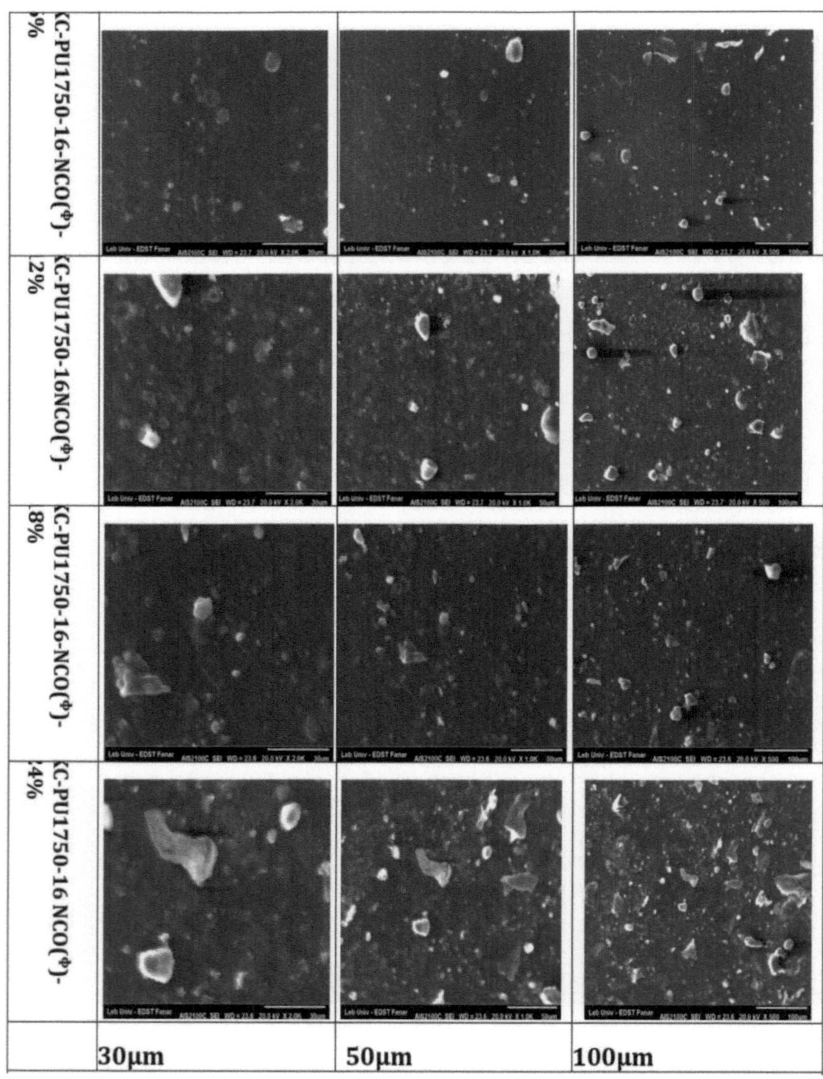

Figure 27-MBE micrographs ofKC1750-16-NCO composites

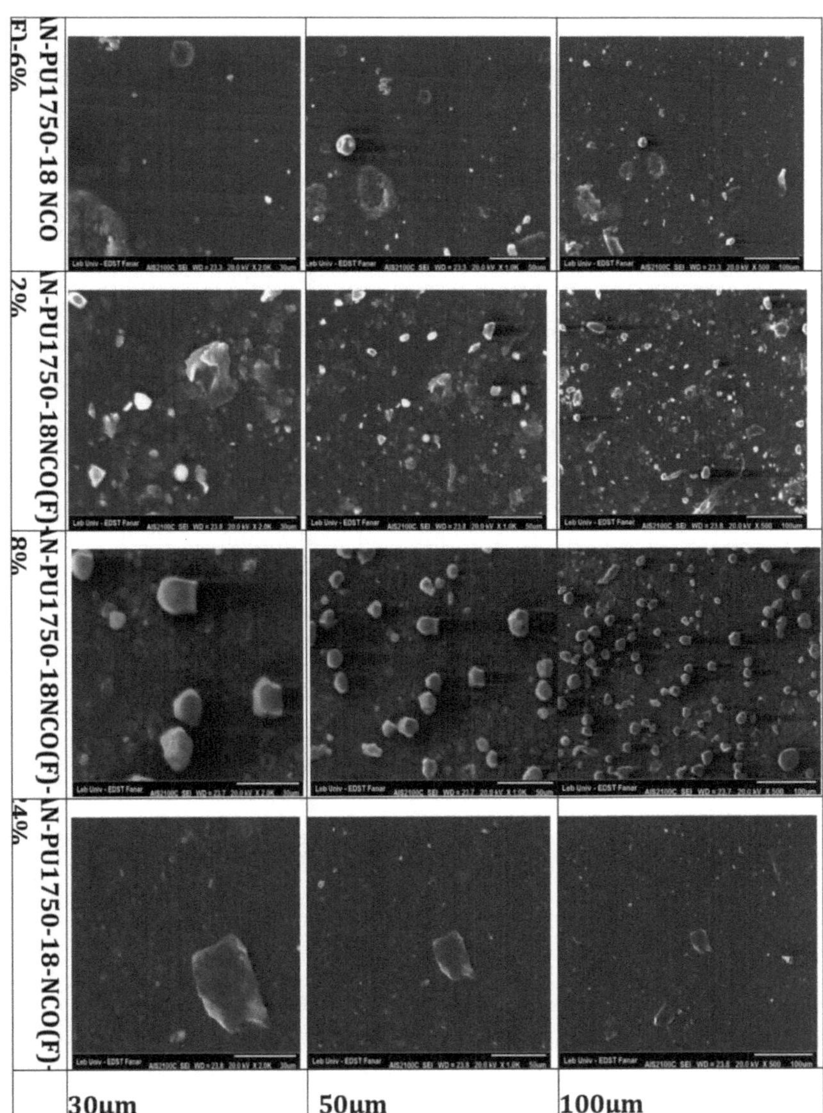

Figure 28-MBE micrographs ofAN1750-16-NCO composites

It seems to us that the interactions established in the parent mixture (polyol-24%NTC-COOH/Citric Acid) are strong, and dispersion, by simple addition of pure polyol, is not evident.

Micrographs of all other composites, KC-PU-1750NTC-COOH/hydroquinone and

KC-PU1200-NTC-COOH/hydroquinone, (FIG-26and FIG-28) show very well-dispersed phases, while composites made with polyol (molar mass M=1200gr) display homogeneous, continuous textures.

2) *Thermal stability by thermogravimetric analysis*

The stability of the various polymers was studied by thermogravimetric analysis, which was carried out on an instrument (Automatic Multiple Sample Thermogravimetric Analyzer TGA-1000-Navas instruments). Tests were carried out at a constant nitrogen flow rate and a temperature rise rate of 10°C/min.

We have defined the temperature at which degradation begins as the temperature at which the sample has lost 10% of its initial mass.

The thermogravimetric degradation profiles of two families, HSPU-X% and HSPU-X%-Si, are shown in figures (FIG-30) and (FIG-31).

Figure 29- Degradation as a function of temperature; PU TGA

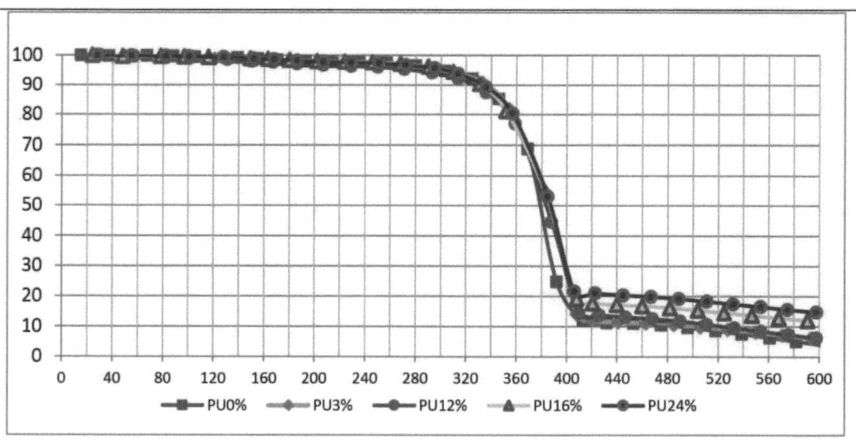

Figure 30-Degradation as a function of temperature.TGA of PU /Si

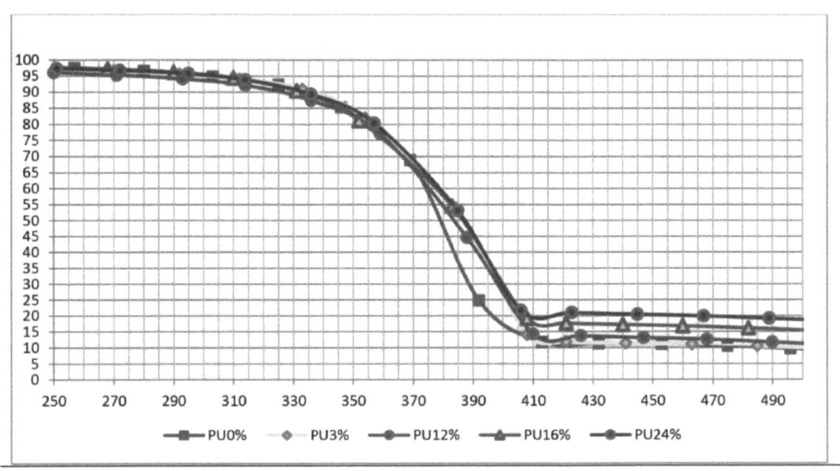

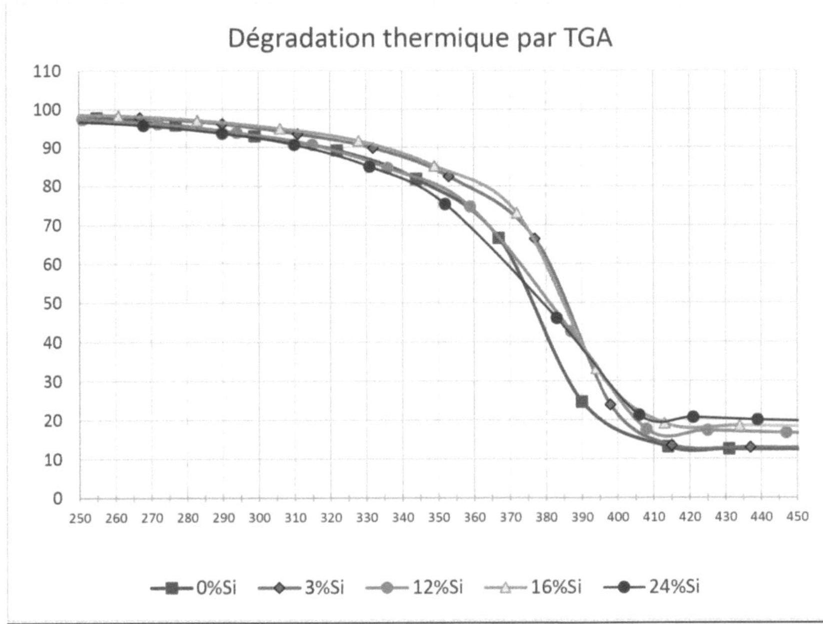

The thermogravimetric degradation profiles of two AN-PU1200-18 NCO- NTC-citric acid families and the KC-PU1200-16-NCO($) NTC-hydroquinone family are shown in figures (FIG-32) and (FIG-33) respectively.

Figure 31- **Degradation profiles of AN-PU-1200 NTC-COOH/Ac. Citric**

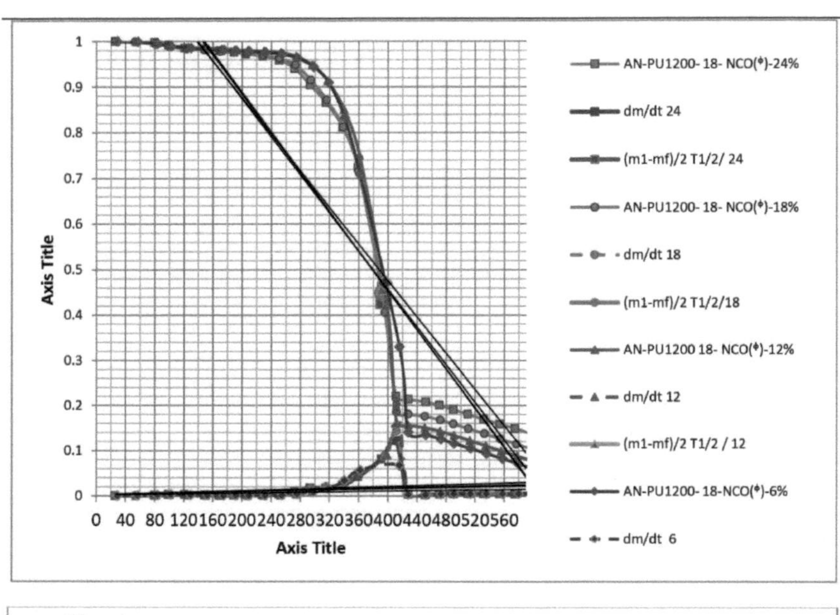

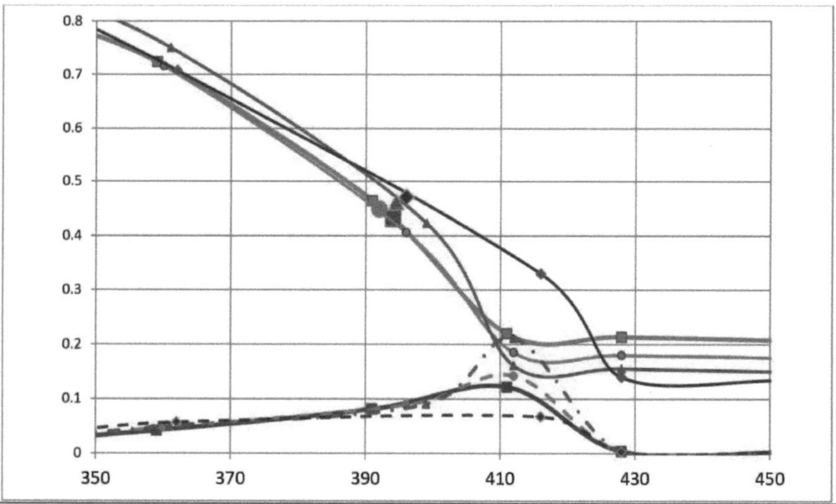

Figure 32-Degradation of **KC-PU1200-NCO-NTC-COOH - hydroquinone Coordinates Enlarged**

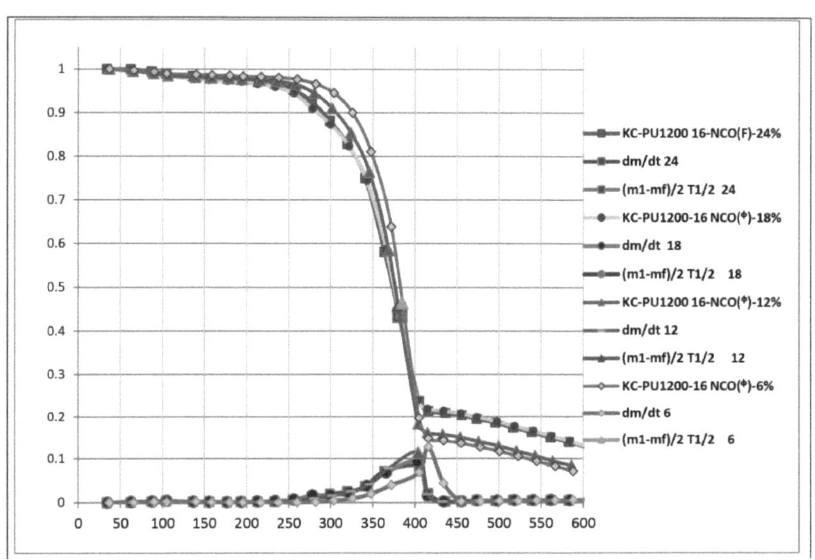

Coordinates Enlarged

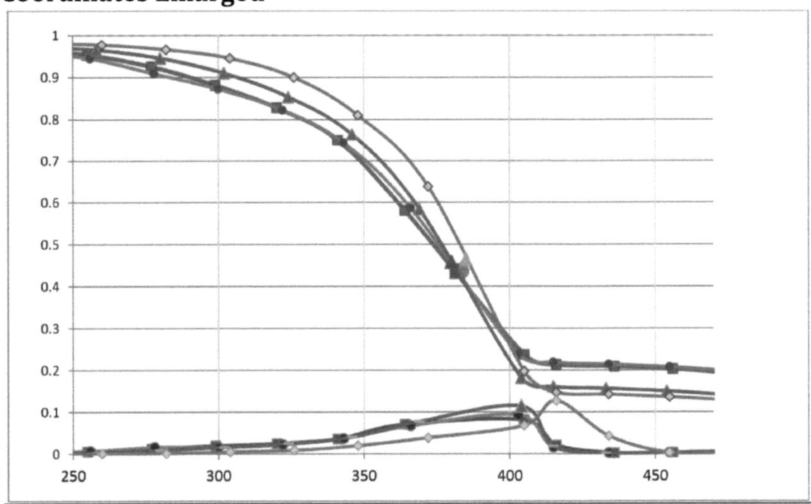

Figure 33-Degradation profiles of **AN-PU-1750NTC-COOH /Ac. Citric**

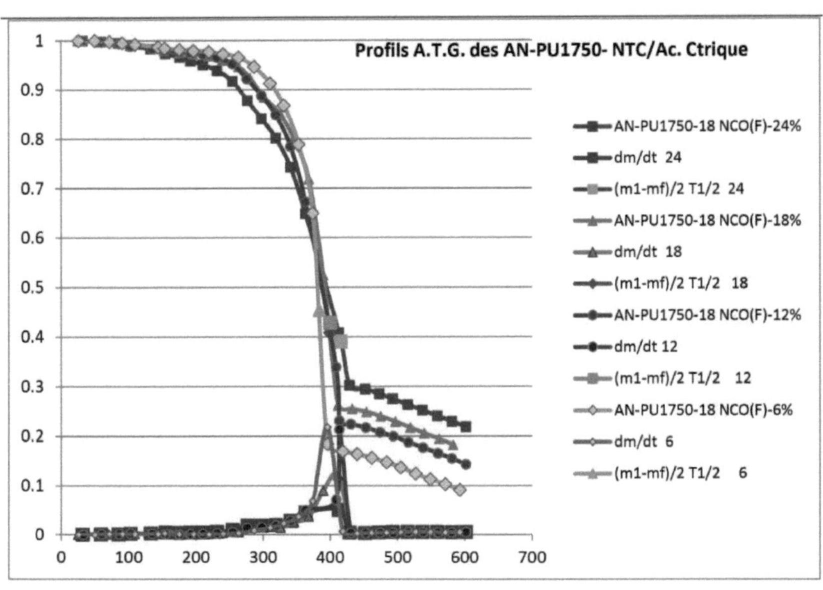

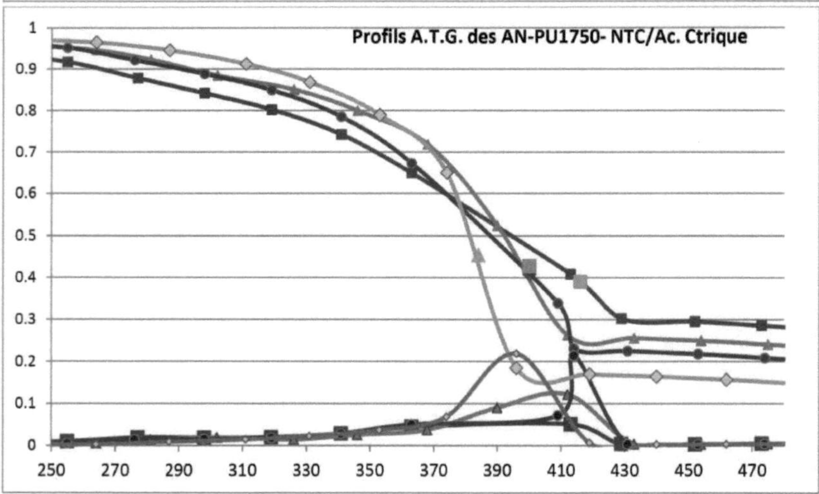

Figure 34- Degradation profiles of KC-PU-1750 NTC- COOH/Hydroquinone

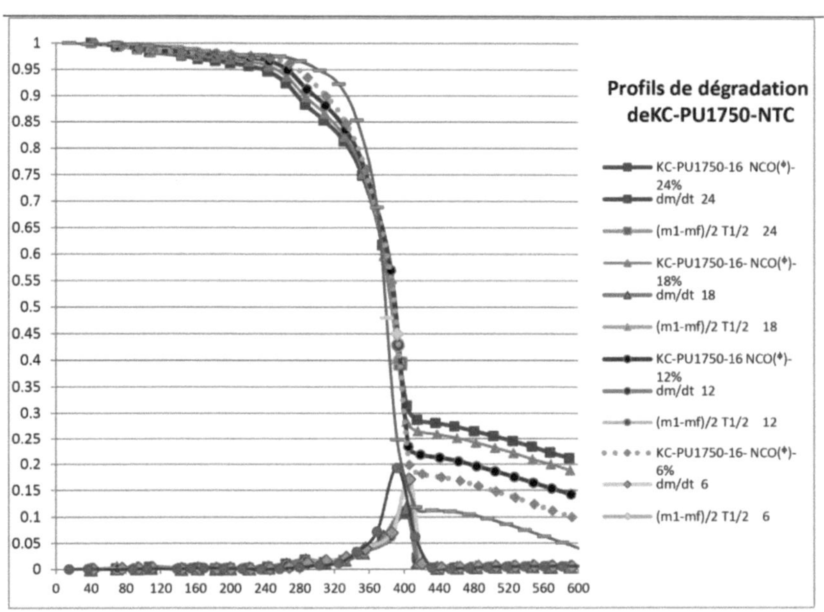

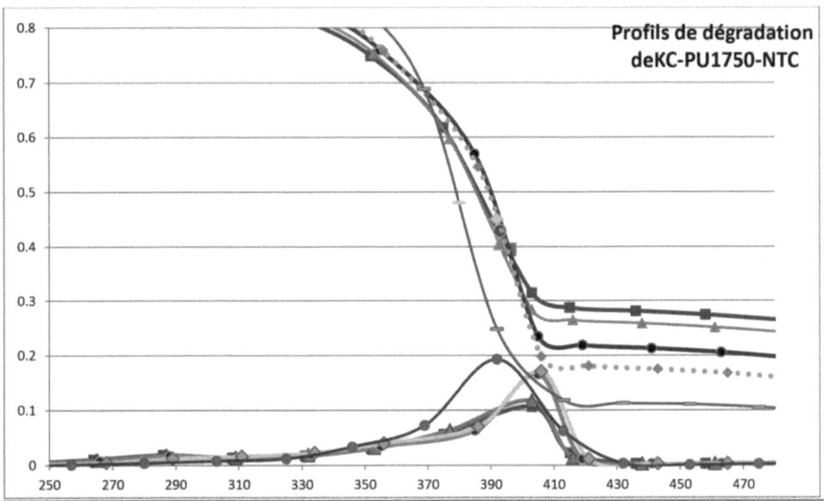

The thermogravimetric degradation profiles of two families AN- PU1750-18 NCO- NTC-citric acid and the KC-PU1750-16-NCO($^\$$) NTC-hydroquinone are shown in figures (FIG-34) and (FIG-35).

For each family, we have plotted the variations in mass and degradation rate of the PUs, with each figure containing two diagrams, one made to normal scale and the other enlarged.

3) *Explanation and commentary on T.G.A. profiles.*

There are four important quantities to consider;

A.   The temperature at which degradation begins

B.   Half-degradation temperature

C.   The temperature corresponding to maximum degradation speeds

D.   The percentage of coal remaining at around 600°C.

A. The temperature at which degradation begins

Samples from the 1ère and 2ème families are said to lose 10% of their mass and begin to degrade at temperatures ranging from {314[0] to 336 C}.[0]

As shown in Table 7, neither the insertion of Si derivatives nor the percentage of carbon nanotubes present in the PU matrix have any appreciable effect on the onset of degradation of PUs manufactured with non-functionalized nanotubes (denoted by an HS prefix). However, for

Table 7-Data extracted from A.T.G. of PU-1750-NTC

PU-NTC-Si (*)	T start	%m	T of half degradation	Maximum degradation speed	TFinale	% in coal
HSPU1750-0%-Si	322	89.22		3900	603	5.288
HSPU1750-3%-Si	332	89.81		3980	591	6.085
HSPU1750-12%-Si	315	90.68		4030	598	9.82
HSPU1750-16%-Si	328	91.72		3950	586	12.97
HSPU1750-24%-Si	314	89.69		4020	591	13.34

PU-NTC /PUR (*)						
HSPU1750-0%	325	92.28	3770	3920	602	3.94
HSPU1750-3%	333	90.89		4060	594	6.23
HSPU1750-12%	314	92.13		4080	598	6.26
HSPU1750-16%	331	90.36		4100	591	12.0224
HSPU1750-24%	336	89.36		4030	598	14.8
PU-NTC / Citric acid (+)						
AN-PU1750-18 NCO (F)-6%	316	0.912	385	396	593	9.12
AN-PU1750-18NCO(F)-12%	298	0.888	376	360-414	602	14.35
AN-PU1750-18NCO(F)-18%	291	0.9	382	407	583	18.27
AN-PU1750-18-NCO(F)-24%	277	0.8779	371	363- 405	602	21.88
PU-NTC/hydroquinone(+)						
KC-PU1750-16-NCO(*)-6%	308	0.9127	386	406	593	10.01
KC-PU1750-16NCO(*)-12%	297	0.9	384	405	592	14.42
KC-PU1750-16-NCO(*)-18%	287	0.8976	377	403	591	18.82
KC-PU1750-16 NCO(*)-24%	275	0.9047	376	403	590	21.18
(+)Functionalized CNTs			(*)Non-functionalized CNTs			

Table 8- Data extracted from A.T.G. of PU-1200-NTC

	T start	%m	T half of degradation	T Maximum degradation speed	TFinale	% in coal
PU-NTC / Citric acid (+)						
AN-PU1200-18 NCO (F)- 6%	32 2	0.9	387	360428	602	5.87

AN-PU1200-18NCO(F)-12%	31 2	0.908	385	360412	600	7.68
AN-PU1200-18NCO(F)-18% FOR THE	29 9	0.902 1	379	360 410	599	1026
AN-PU1200-18-NCO(F)-24%	29 3	0.904	378	360 408	597	13.7
PU-CNT/hydroquinone(+)						
KC-PU1200-16-NCO(*)-6%	32 2	0.9	380	405	588	7.05
KC-PU1200-16NCO(*)- 12% FOR THE	30 2	0.902 1	373	401	586	8.38
KC-PU1200-16-NCO(*)-18% FOR THE	28 6	0.908	368	399	601	13.0
KC-PU1200-16 NCO(*)-24% FOR THE	28 0	0.905	366	396	601	12.79
(+)Functionalized CNTs			(*)Non-functionalized CNTs			

Figure 35-Variation of temperature at onset of degradation with % CNT

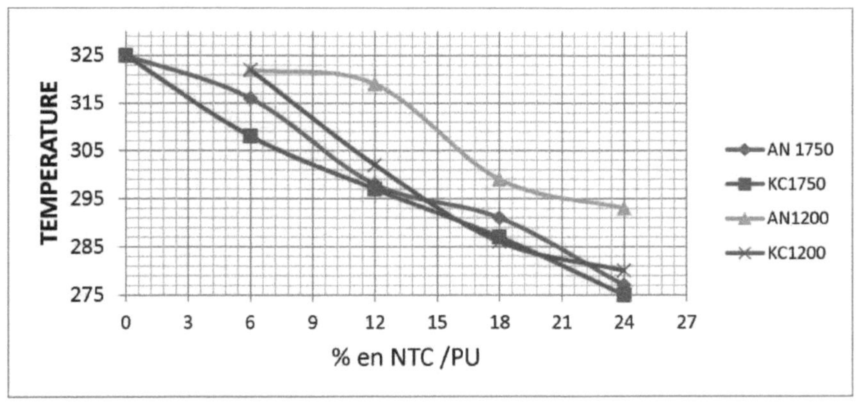

B. Half-degradation temperature

The half-degradation temperature determined from thermogravimetric analysis data corresponds to the loss of half the degassed mass (see chapter -4- for its determination). The values of this parameter for composites from four PU-NTC families show the same trend already seen in the case of the start of degradation: increasing the rate of incorporation of CNTs leads to a decrease in the values of the half-degradation temperature, as carbon nanotubes facilitate the thermal degradation of the materials examined. Figure (FIG-37) shows the variation in the half-degradation temperature of PU-CNTs as a function of the mass percentages of functionalized CNTs.

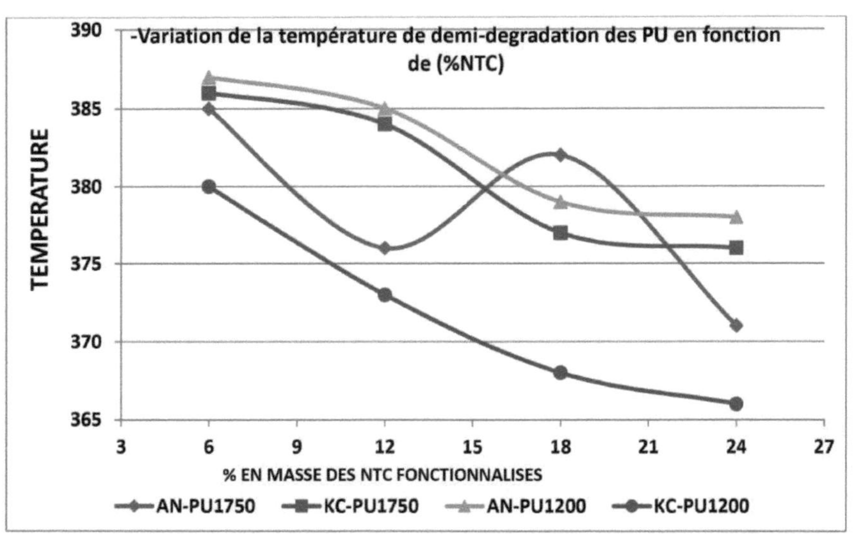

Figure 36 -Variation of PU half-degradation temperature as a function of (%NTC)

A[23] group of researchers has published that both pure PU foam and its nanocomposites decompose in a two-stage process, with the ATG curve of the PU composite filled with 0.5% by weight carbon nanotubes shifting to a higher temperature than that of pure PU foam, as shown in figure (FIG-38). A comparison of the ATG curves for the two foams shows that all the degradation temperatures corresponding to losses of 1%, 5% and 50% by weight are higher for the filled PU than for the pure PU foam, in particular for a loss of 50% by weight, T rises from 450 to 499⁰ C. This indicates that the incorporation of carbon nanotubes at such a low concentration induces a remarkable thermal stabilization of the matrix. The results also revealed that the uniform, fine dispersion of CNTs improved the interfacial adhesion between the[23] CNT and the matrix.

TABLE 11

Degradation Properties of CNTs/PU Foam Composites

Foam type	1%	5%	50%
PU	220	275	450
PU: 0.5 wt % CNTs	237	281	499

[23]

This result can be obtained in our work if we are satisfied to compare only three samples:

HSPU1750-0%	3770
AN-PU1750-18 NCO (F)-6%	**385**
KC-PU1750-16-NCO(*)-6%	**386**

We obtained different results when the CNT load increased beyond 6% to 12, 18 and 24%.

C. The percentage of coal around 600°C.

However, the mass of residual char at the end of degradation in all samples from all families increased with increasing carbon nanotube content. This observation supports the idea that carbon nanotubes can act as flame retardants. Tables (8) and (9) show the final degradation temperatures and remaining coal masses.

H. Xia & M. Song[5..p=393] reported that the thermal degradation procedure for polyurethanes is complicated and occurs in two stages;

• the initial degradation of step (1) is mainly the decomposition of hard segments, which involves the dissociation of the urethane function into the original polyol and isocyanates, and which then forms a primary amine, an alkene and carbon dioxide, this step is influenced by the content of the hard segment.

• Step (2), which follows, takes place according to a depolycondensation and polyol degradation mechanism, and is affected by the contents of the flexible segment.

In general, the temperature corresponding to the maximum mass loss rate in the TGA curve is accepted as the thermal decomposition temperature of the composite. This group found {for *a PU formed by 1- a polyol (M = 6000) - CNT in dispersion, 2- MDI isocyanate, 3- Extender; 1,4-butanediol)* that the 1st degradation of unfilled polyurethanes occurs at 332^0 C and the second degradation at 393^0 C. The incorporation of MWNT and SWNT nanotubes did not improve the 1ère degradation temperature, but it did improve the second degradation temperature. This indicates that carbon nanotubes can preferentially interact with the flexible segment in the polyurethane structure.

For a PU loaded with nanotubes by 1% of its mass, the second degradation temperature will be delayed by 5 C.^{0}

The TGA curve[6..p=3804] shows a dramatic drop in the temperature range 300^0 C to 400^0 C, due to the degradation of the rigid segments, which had weak bonds in the PU-SWCNT hybrid. These rigid segments would have been broken and produced free radicals, which would attack the flexible PU-SWCNT hybrid segments, resulting in a chain reaction of continuous ruptures.

D. Degradation speed profiles

To explain and comment on the degradation rate profile data, we have classified the samples into three categories according to the CNT processing mode:

a) The degradation rate profiles of PU samples whose carbon nanotubes have not been functionalized are shown in figures **(FIG-38)** and **(FIG-39),** each profile presenting a maximum degradation rate at a given temperature. It can be seen that (with the exception of PU24%) this speed gradually decreases as the percentage of carbon nanotubes increases, and that the temperatures corresponding to the peak maxima also increase as the percentage of nanotubes increases;

Figure 37- Degradation rate of PUs from the HSPU1750- X%C family

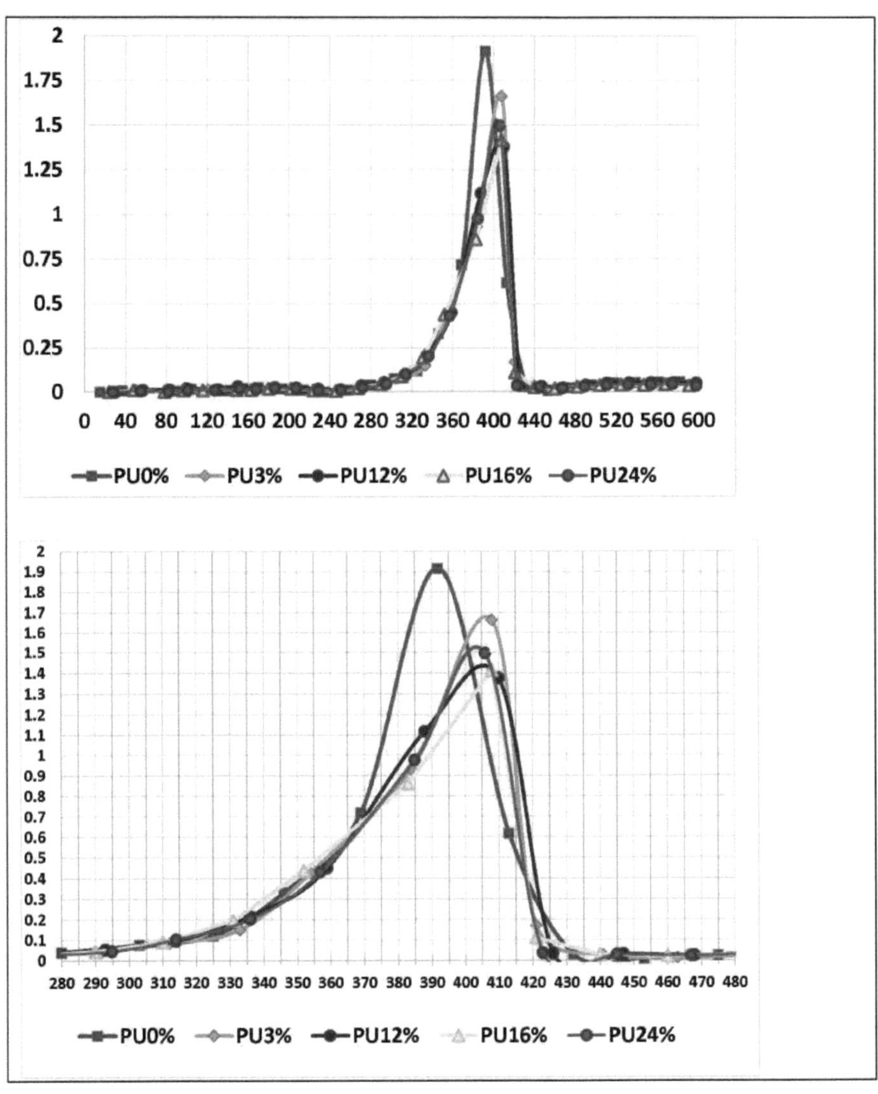

Figure 38- Degradation rate of PUs from the HSPU1750- X%C Si family

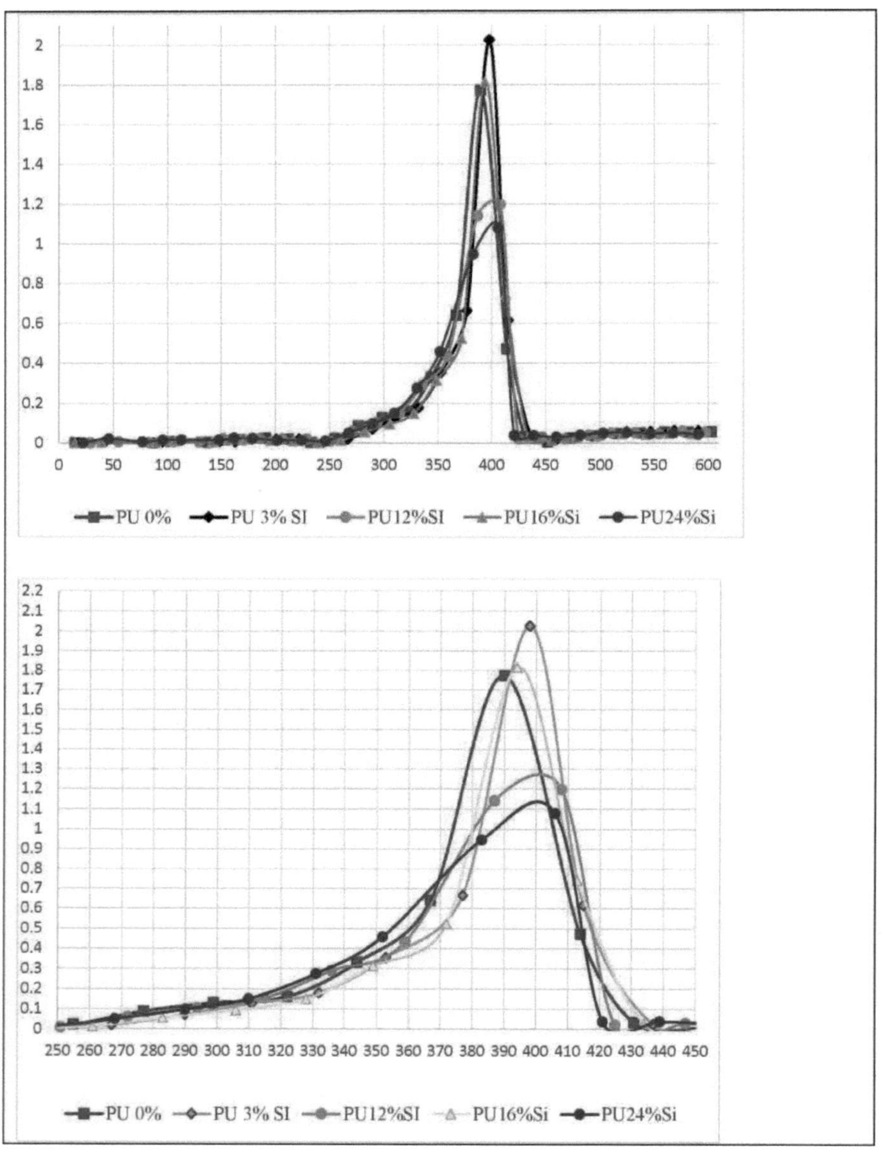

Generally speaking, we can conclude that the introduction of carbon nanotubes increases the material's thermal stability;

Table 9 - Coordinates (speeds, temperatures) of the maximum points of the thermogravimetric profile peaks of PU and PUSi

$V_{PU0\%} = 1.91$ >	$V_{PU3\%} = 1.66$ >	$V_{PU12\%} = 1.44$ >	$V_{PU16\%} = 1.38$	$V_{PU24\%} = 1.52$
T=392⁰ C	T=406⁰ C	T=408⁰ C	T=410⁰ C	T=403⁰ C
$V_{PU0\%Si} = 1.78$	$V_{PU3\%Si} = 2.05$	$V_{PU12\%Si} = 1.27$	$V_{PU16\%si} = 1.82$	$V_{PU24\%Si} = 1.12$
T=390⁰ C	T=398⁰ C	T=403⁰ C	T=395⁰ C	T=402⁰ C

This increase in stability can be expressed either by a decrease in degradation speed or by the deviation of the maximum point of the degradation peak towards slightly higher temperatures. The insertion of Si, on the other hand, reduces stability.

Degradation rates are compared between HSPU-1750-0.0% and HSPU-1750-0.0%-Si (FIG-40) and between PU-16% and PU-16%-Si (FIG-41).

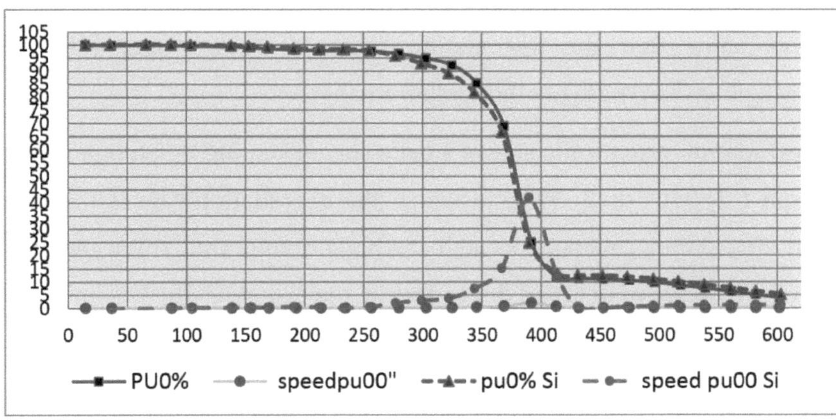

Figure 39-Comparison of degradation rates between PU 0.0% and PU0.0%-Si

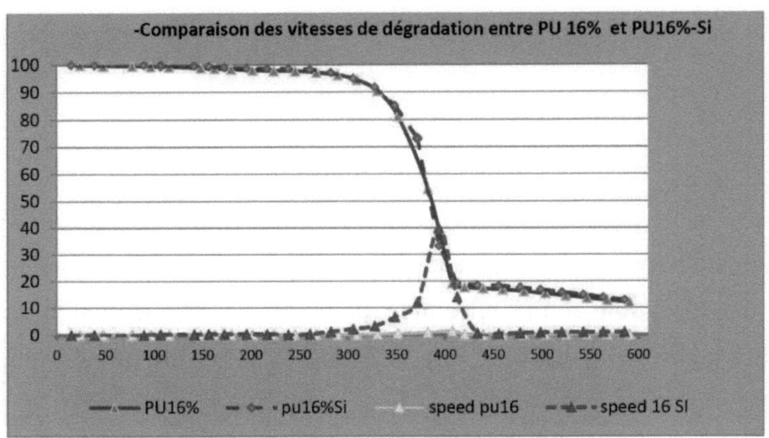

Figure 40-Comparison of degradation rates between PU16% and PU16%-Si

Both figures show that the degradation rates of Si-containing PUs are significantly higher than those of other polyurethanes.

b) Each degradation rate profile shows a belly corresponding to a temperature interval from { T=360⁰ to T~395}, in which the degradation rate is average, followed by a maximum rate peak located at a higher temperature. We consider this belly to be **a phase of delayed degradation**, the extent of which is linked to the quantity of interactions between the chains. In our case, these interactions are proportional to the number of NTC-COOH functions. This estimate will be more credible when we notice the total or partial relative absence of this belly in non-functionalized PU-NTC- composites. The profiles of composites made with oxidized CNTs then treated with citric acid (ANPU....NCO...NTC-COOH- Citric acid series) are shown in the figure **(FIG-42).** With the exception of

Figure 41- Degradation rate profiles of AN-PU-poyol-NTC- COOH/citric acid

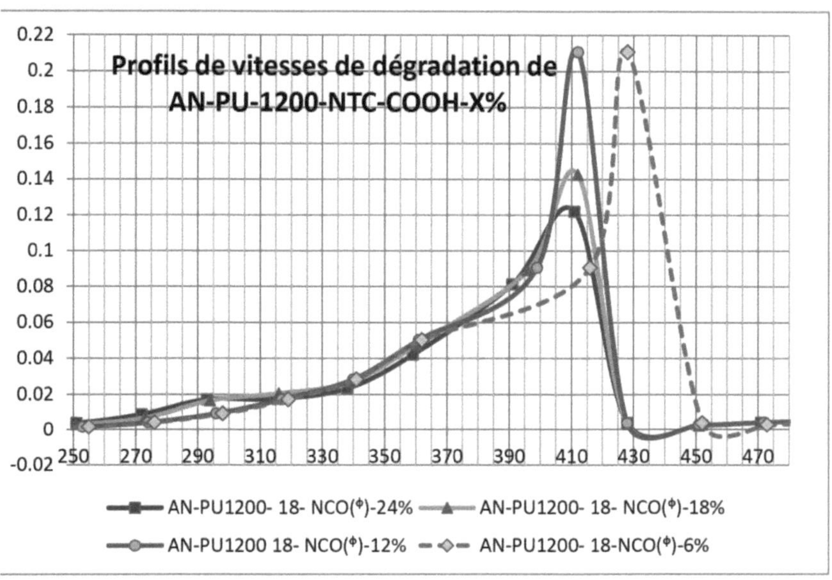

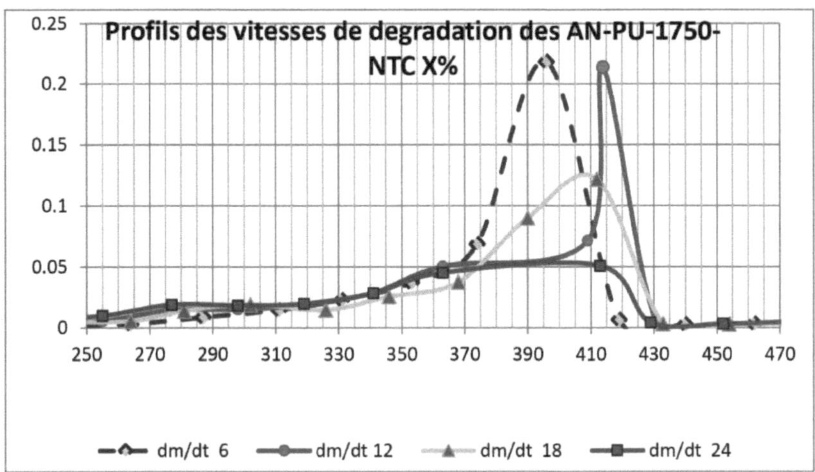

With the non-homogeneous sample {AN-PU1200-18 **NCO (F)-6%}** (see micrograph -FIG-27, page 52), which decomposes at the highest rate and at a higher temperature than the other PU- CNTs, we can conclude that the material degradation rate is a function of the mass percentages of nano-carbons. The effect of CNT-COOH as a thermal stabilizer is best demonstrated on the profile of **ANPU-1750-NCO-18-NTC-COOH-24%,** a material with a high concentration of

CNT-COOH, which decomposes over a wide temperature range, at a low and almost constant rate, in the absence of a peak corresponding to a maximum rate.

In the ATG curves, for a series of PUs similar[12] to our products and synthesized from HMDI + Polyol : PTMG **(M= 1000), the** temperature peaks, which correspond to the maximum degradation rate for the rigid and flexible domains of the treated PU matrix, are between 321 and 408⁰ C, which are approximately improved, for the TPU nano-composites filled with 2.5% by weight COOH-MWNT, to 347 and 418 C.⁰

Figure 42- PUs Synthesized from HMDI + Polyol: PTMG (M= 1000)

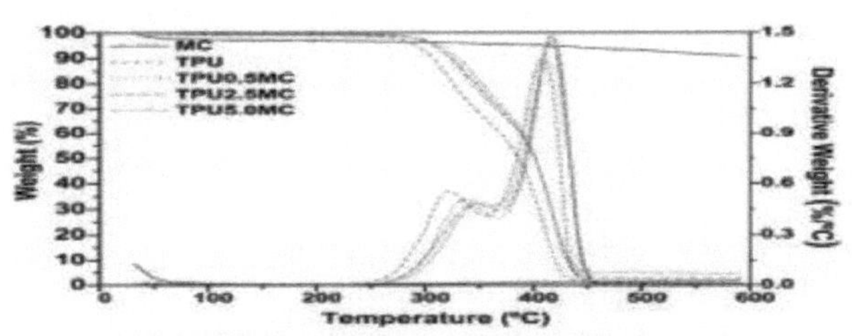

Figure 43- Degradation rate profiles for KC-PU-polyol-NTC-COOH/ Hydroquinone

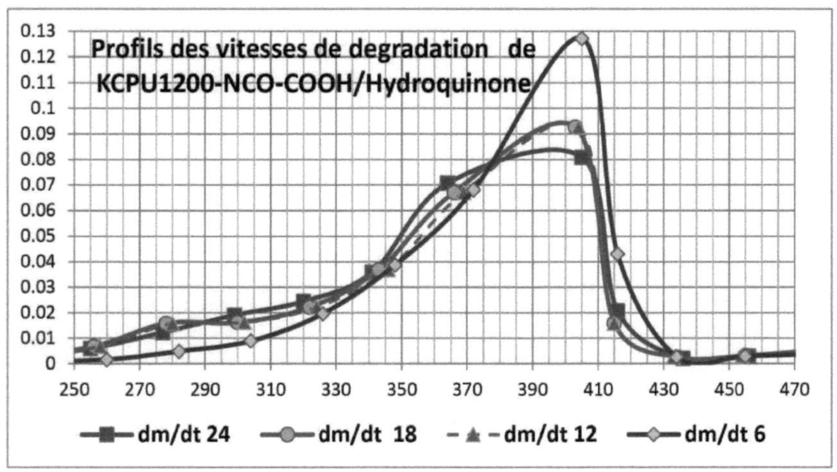

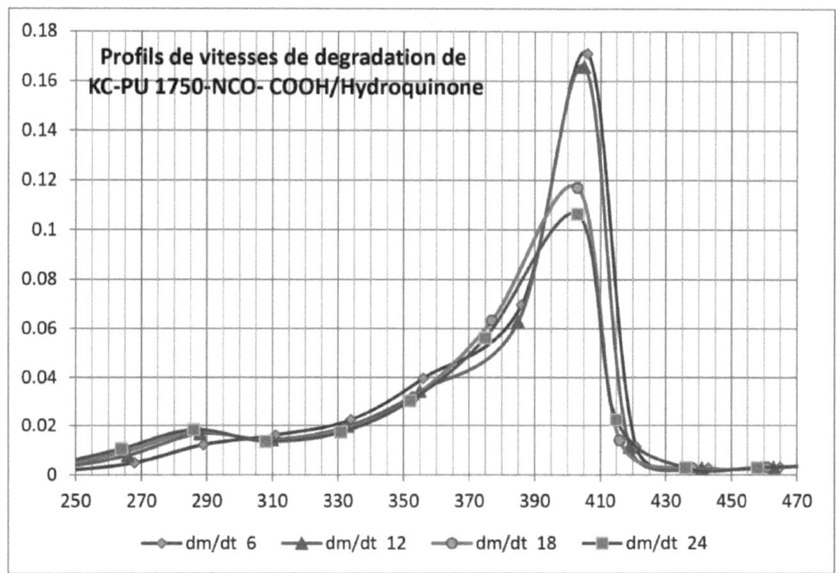

Figure 44- Degradation rate profiles for KC-PU-polyol-NTC- COOH/ Hydroquinone

c) The profiles of hydroquinone-treated PU-polyol-NTC-COOH composites are shown in figure (FIG-43), the relationship between the maximum degradation rate of each sample and its loading rate with functionalized nanotubes is clearly generalized for all composites, this rate decreasing gradually with increasing percentage of functionalized nanotubes, The increase in thermal stability is caused by the presence of thermally stable carbon nanotubes and independently of the molar mass of incorporated polyol.

4) Breaking strengths and elongations

The mechanical properties of a[10] nanoscale reinforced polymer depend on several factors such as the distribution and orientation of the nanofillers, their size ratios, and their degrees of compatibility [28].

In the case of a PU-CNT composite, the properties are related to all the forces and interactions (Vander VALS, H-bonds,..., covalent bonds, ionic bonds), the distribution, separation and size of the material's hard and flexible micro-phases, as well as its thermomechanical history and shape dimensions.

MWNT and SWNT nanotubes have walls consisting of polyaromatic structures with π-conjugated bonds, which can be retained by electrophilic sites existing, possibly, in the polymer matrix. The introduction of -COOH on the MWCNT surface wall facilitates hydrogen bonding with functional groups (urethane), while additional ester and carboxylic COOH groups can also contribute to the good interaction between CNTs and PU.

The considerable improvement in the mechanical properties of composite materials is not linked to a simple quantification of molecular interactions, but to other more important parameters, such as a homogeneous dispersion of CNTs throughout the polymer matrix and strong interfacial adhesion between oxidized or esterified nanotubes and the matrix.

Tensile strength and elongation at break are well related to the interactions between functionalized nanotubes and macromolecular chains in the flexible phase of the composite, i.e., elongation corresponds to the average elongation of expanding chains between two nodes, which can arise from a number of sources, such as covalent bridging, chain entanglement and local crystallization prompted by nanotube aggregates, which can promote nucleation and act as physical cross-linking points, limiting the movement of polymer chains; localized crystallization prompted by the formation of nanotube aggregates, which can promote nucleation and act as physical cross-linking points, limiting the movement of polymer chains.

The tensile tests were carried out using two machines: the first was used to cut the specimens to be examined, according to a well-defined shape and dimensions (see Chapter 4), the other to carry out the tensile tests.

Tensile testing is carried out at room temperature with an elongation rate of 25mm/minute. From each sample, two specimens are cut and examined, and an average of two values is reported.

A. The samples are classified into two categories; the first contains the two families [AN-PU1750-18- NCO...%] and [AN-PU1200-18- NCO...%], which are a kind of polyurethane containing nanotubes with COOH functions, ester bonds, phases connected by hydrogen bonds and urethane bonds and probably by hydrogen bond bridges, between the ester groups or -COOH functions of the oxidized carbon nanotubes on the one hand and the -OH functions of the flexible polyol segments on the other.

The results are shown in tables (Table-10) and (Table-11) and are as follows

Table 10- Tensile test; average elongation and percent elongation and breaking strength for PU prepared with MDCI and polyol (M=1200, f=3) and oxidized, citric acid-treated nanotubes.

	Alongement Al average	% average elongation	Average force applied
AN-PU1200-20-NCO(à)-24%-A	166	664	2
AN-PU1200-18-NCO(*)-24%-A	193.375	773.5	2.5
AN-PU1200-18-NCO(*)-18%-A	97.25	388.7	4
AN-PU1200-18-NCO(*)-12%-A	61.9	258.4	5.5
AN-PU1200 -18-NCO(*)- 06%-A	51.2	206.4	6.5

show for each sample the values for; elongation, percentage elongation and force applied at break.

Table 11- Tensile test; average elongation and breaking strength for PU prepared with MDCI and polyol (M=1750, f=3) and oxidized, citric acid-treated nanotubes.

	Elongation average Δl	% elongation average		Force applied
AN-PU1750-18-NCO(*)-24%-A	43	171.6	10	
AN-PU1750-18-NCO(*)-18%-A	93.45	373.8	7	
AN-PU1750-18-NCO(*)-12%-A	87.4	349.6	8	
AN-PU1750-18-NCO(*)- 06%-A	72.8	291.2	10.5	

Figure 45-Traction **test of** AN-PU 1200-18- NCO-NTC-citric acid Variation of Δl and % elongation with % C

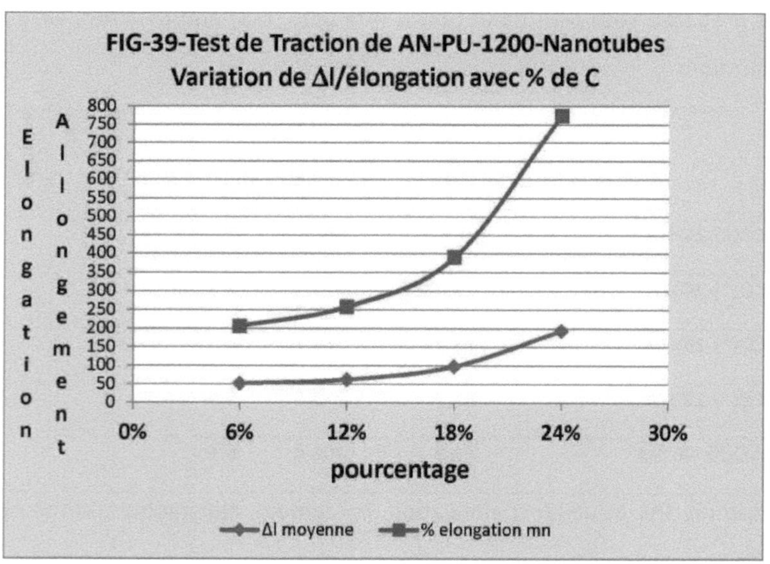

For composites made with polyol (M=1200gr), elongation and elongation at break are both functions of the CNT content in the composites. The elongation of the composite loaded with 24% CNTs, relative to the mass of polyol (10% relative to the mass of PU) is of the order of 773%, a behavior close to that of an elastomer, i.e. 3.75 times the elongation of the composite loaded with 6% CNTs. Fracture strength expressed by the force applied at the breaking point is inversely proportional to the percentage of CNTs, with the composite least loaded with carbon nanotubes showing the highest fracture strength. A material that has undergone significant elongation has a lower breaking strength than one that has been slightly elongated.

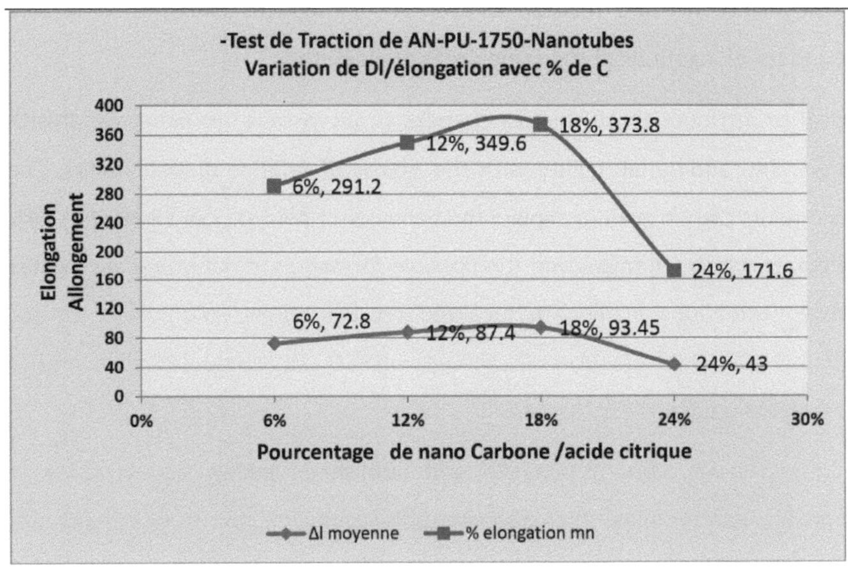

Figure 46- **AN-PU 1750-18- NCO-NTC-citric acid tensile test**

Variation of Δl and % elongation with % C

Composites made with (M=1750) polyol are characterized by three changes compared to composites made with (M=1200) polyol;

They have longer chain lengths, i.e. a smaller number of chains, a higher

$$\frac{masse\ des\ NTC-Fonctionnalisés}{nombre\ des\ chaines}$$

ratio, i.e. more frequent nanotube-chain

interactions, and a higher $\frac{NCO}{POLYOL}$ ratio; i.e. a higher percentage of hard units. A higher percentage of hard units.

Table -12 shows, with the exception of the most loaded composites (24% CNTs), the same trends already observed above, for elongation and elongation, but with a lower slope than the previous composites. A comparison of the values of the forces applied at fracture between the two families reveals higher values for the 2ème family, and higher fracture resistance, even for composites with the same CNT content and for elongations of the same order of magnitude.

It is logical to attribute these relative changes to an increase in nanotube-sheath interactions and additional rigidity with the excess of relative di-isocyanates. The same arguments can be used to explain the behavior of **AN-PU1750-18-NCO(*)-24%** composites, to which we might add the possible formation of nanotube aggregates that cause nucleations leading to localized crystallizations, and which can be revealed by scanning electron microscopy (SEM) of composites at higher concentration levels (see FIG- 28 page 71).

Similar results have been obtained[1,9] and published, linking the decrease in elongation at break for a high CNT concentration such as 17.7 wt.%, to the increase in the frequency of localized aggregation.

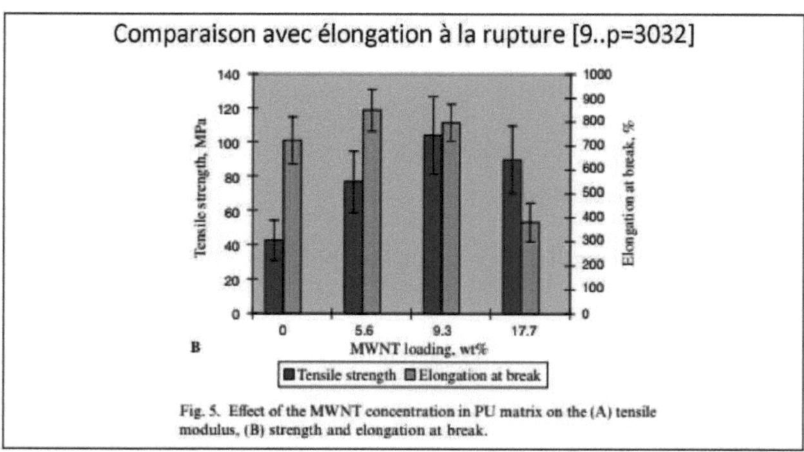

Fig. 5. Effect of the MWNT concentration in PU matrix on the (A) tensile modulus, (B) strength and elongation at break.

The results show that the interfacial interaction between the TPU matrix and CNT-COO-{COOH}2 increases with increasing %wt CNTs.

B. The 2ème category includes the two families [KC-PU1200-16- NCO...%] and [KC-

PU1750-16- NCO.%], which are a kind of polyurethane containing nanotubes with fewer COOH functions than the composites of 1ère category, the addition of hydroquinone would not result in a quantitative esterification of phenolic hydroxyls, a possible and partial esterification may occur between the COOH functions grafted on the nanotubes, on the one hand, and the polyol hydroxyl functions. The di-isocyanates then added will react with both the polyol hydroxyl groups and the phenolic OH, and the hydroquinone will be incorporated into the composite matrix.

The ratio was increased **from** **to** $\frac{25}{18}$) $\left(\frac{NCO}{Polyol} = \frac{25gr}{16} \right.$ to compensate for hydroquinone's participation in polycondensation.

The results are given in Tables (Table-12) and (Table-13), which show for each sample the values; of elongation, percentage elongation and force applied at break.

Table 12- Tensile test; average elongation percentages and average elongations at break for PU prepared with MDCI and polyol (M=1200, f=3) and hydroquinone-treated nanotubes

	Δl average	% elongation mn	Force applied
KC-PU1200-16- NCO(*)-24%-A	268.7	1074.8	2
KC-PU1200-16- NCO(*)-18%-A	84.15	336.6	5
KC-PU1200-16- NCO(*)-12%-A	72.15	288.6	6.5
KC-PU1200-16- NCO(*)-6%-A	52.5	210	7

Figure 47-Tensile testing of KC-PU-1200-Nanotubes Variation in Dl/elongation with % C treated with hydroquinone,and comparison of tensile testing between ANPU1200-18-NTC-COO(COOH)₂ and KC1200-16-NTC-COOH- Hydroquinone composites.

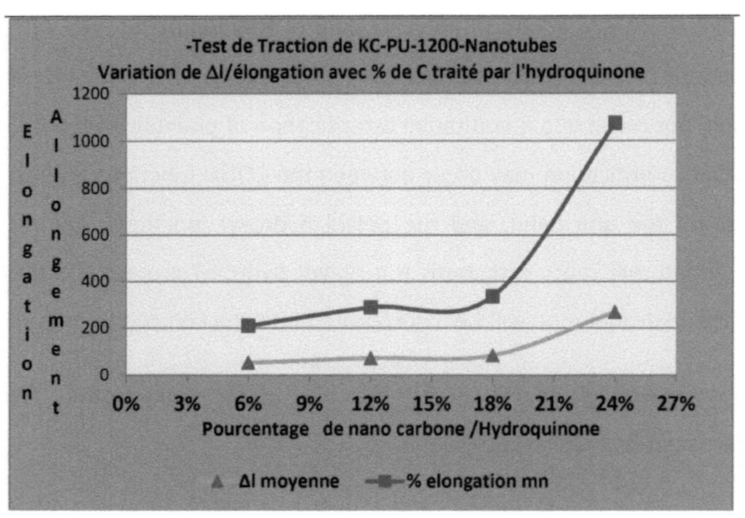

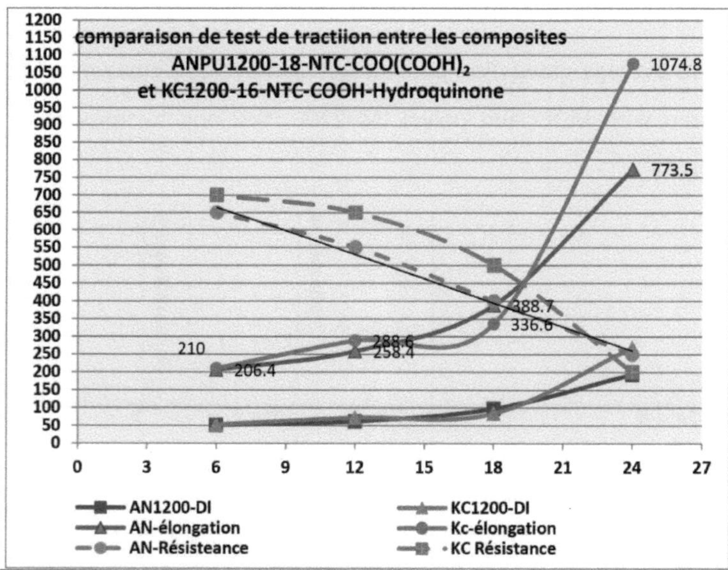

Figure (FIG-47-) shows variations in behavior identical to those already seen in the ANPU1200-18-NCO-NTC-COOH/citric acid family.

Elongation values are slightly higher for composites treated with hydroquinone, especially the sample containing 24% CNTs which shows elastomer-like elongation (% elongation>1000%), as well as fracture toughness values.

For the last family made with a polyol of molar mass (M=1750), a linear relationship can be established between both elongation and percentage elongation at break on the one hand, and the percentage loading of functionalized CNTs on the other. Table (-13-) and figure (-48-) show these results.

Table 13- Tensile test; average elongation percentages and average elongations at break for PU prepared with MDCI and polyol (M=1750, f=3) and hydroquinone-treated nanotubes

	Δl average	average % elongation	Force applied
KC-PU1750-16- NCO(*)-24%-A	151.35	605.65	6.5
KC-PU1750-16- NCO(*)-18%-A	127.15	508.5	7
KC-PU1750-16- NCO(*)-12%-A	95.55	382.5	8
KC-PU1750-16- NCO(*)-6%-A	68.15	272.6	10.5

The high fracture toughness values of all the composites correspond to low CNT filler contents, and at the same time to low fracture elongation values, and the converse is true: a comparison of the fracture toughness of all the samples shows that the type of nanotube treatment has no influence on the fracture toughness of the composites, but this quantity is dependent on the polyol size.

FIG-48-Tensile test of KC-PU-1750-Nanotubes Variation of Δl/ Percent elongation/ Carbon nanotube/Hydroquinone

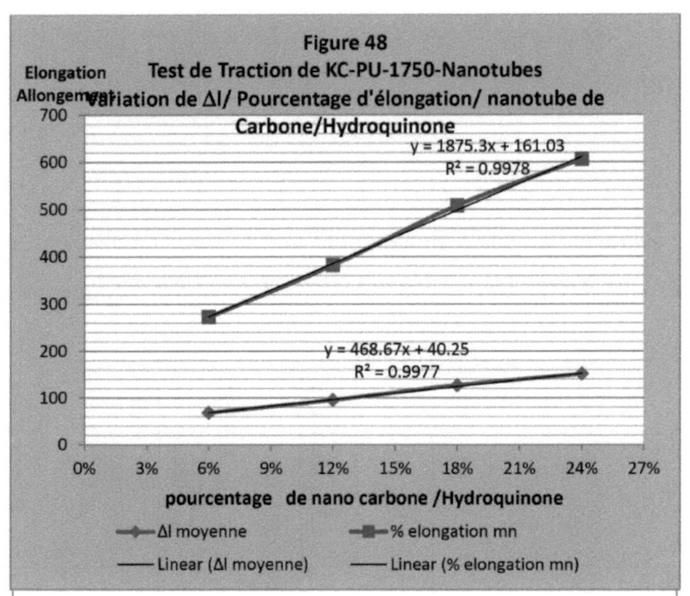

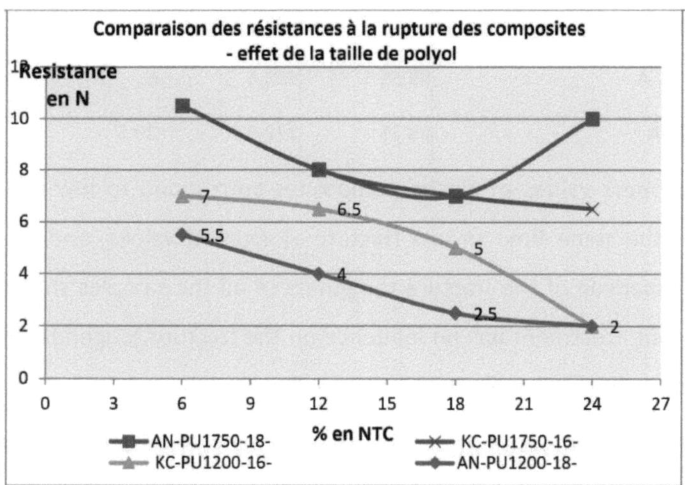

The slightly higher fracture toughness values of the [KC PU 1200-16NCO] series compared to the AN PU 1200-16NCO series can be attributed to the insertion of phenol nuclei in the M=1200gr polyol chains, an effect not observed in the case of the M=1750 polyol.

The linearity of the relationship between elongation at break and CNT filler content in a PU-CNT composite has been published, by another group[16] , in their case elongation is inversely proportional to CNT mass percentages for filler composites varying from 0 to 7% by weight of material.

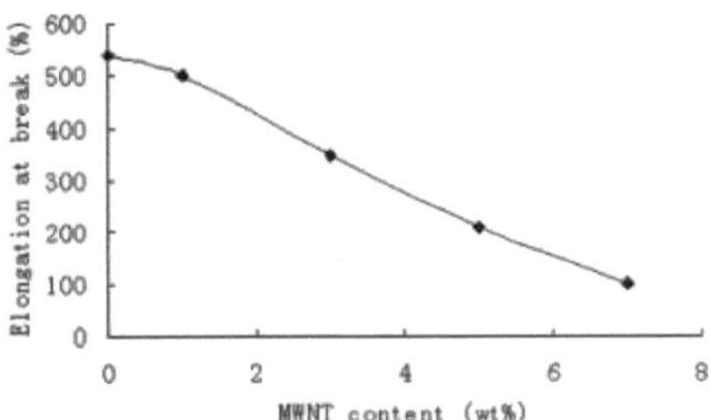

Figure 15 Relationship between the elongation at break of the SMP–MWNT fibers and the MWNT content.

[16]

CHAPTER 4. EXPERIMENTAL TECHNIQUES

1.   Reagents and materials :

I Dicyclohexylmethane 4,4'-diisocyanate MDCI with molar mass 262gr and functionality f =1.43 is supplied by DOW chemicals.

II   The two polyether polyols with molar masses of 1200gr and 1750gr respectively and the same functionality f = 3 are industrial polyols supplied by a Chinese company.

OHCH2 CH2 CH2 $CHOH-[-(CH_2)_4O-]_n-CH2$ CH2 CH2 $CH2OH$

III Tin dibutyl laurate; D.B.T.L. ($C_{32}H_{64}O_4Sn$) is supplied by (Aldrich).

IV The carbon nanotubes are supplied by a Japanese chemist as a courtesy and collaboration.

V   Trimethoxyphenylsilane $C_6H_5Si(OCH_3)_3$ (molar mass: 198.29 gr) supplied by sigma Aldrich with formula ;

VI Citric acid or 2-hydroxy-1,2,3-propane-tri-oic acid (structural formula: $(HOOCCH_2)_2C(OH)COOH$),

VII .hydroquinone $C_6H_6O_2$ (M=110g/mol) or benzene-1,4-diol

HO—⟨ ⟩—OH

2. Synthesis, reactions and application

We have prepared three types of masterbatches with a high carbon nanotube concentration of 24% of the polyol mass:

- *type I masterbatch*; This type of masterbatch is a simple physical mixture, with a carbon nanotube concentration of 24% relative to the mass of polyol. It was prepared by adding 84g of polyether polyol (molar mass 1750g or 1200 and functionality f = 3) to 20.16g of carbon nanotubes and subjecting it to cold agitation for 40 minutes.

- *type II parent mixture*; characterized by functionalization of carbon nanotubes with a sulfo-nitric mixture followed by esterification of carboxylic group COOH with citric acid hydroxyl:

a) Carbon nanotube oxidation is carried out by adding 42g of carbon nanotubes to 10ml H_2SO_{4pur} sulfuric acid, i.e. a mass m=17.6g (M=98g/mol or n=0.179mol) and 4ml pure HNO_3 nitric acid, i.e. a mass m=4.77g (M=56 g/mol or n=0.085mol) n $_{H+}$ = 2 n H_2SO_4 = 0.358 ion/mol.

Assuming that the order of magnitude of the mass of a nanotube varies between 1000 and 1600 u.m.a., the quantity of nanotubes varies between (0.026 and 0.042 mole nanotubes).

The reagents are added gradually, bearing in mind that oxidation is exothermic, and the mixture is placed in a beaker and stirred cold for 2 hours.

b) Esterification of the carboxyl group COOH with the hydroxyl group of citric acid

To the mixture of oxidized carbon nanotubes (approx. m= 62gr) is added a mass m=8.66g of citric acid $C_6H_8O_7$ (HOOCCH$_2$)$_2$C(OH)COOH); (M=193g/mol) , i.e. citric **acid** =0.04487 mol. Nmole OH= (1/8) n $_{H+}$

The beaker is placed in a water bath containing glycerol and the mixture is heated to 120°C for 3 hours to promote esterification.

c) Addition of polyether polyols

From two quantities (approx. 2x71gr) of carbon nanotubes oxidized and esterified with citric acid, two masterbatches containing 24% by weight of functionalized nanotubes are prepared, one with polyether polyol (M=1200gr, f=3) and the other with polyether polyol (M=1750gr, f=3). 133g of polyol are added, the temperature is set at 80°C and the mixture is left to stir mechanically at a speed of 500 rpm. After 24 hours, the polyol masterbatch containing 24% by weight of carbon nanotubes is obtained. Treatment with citric acid increases or even doubles the amount of COOH bound to the nanotube walls, according to the following reaction;

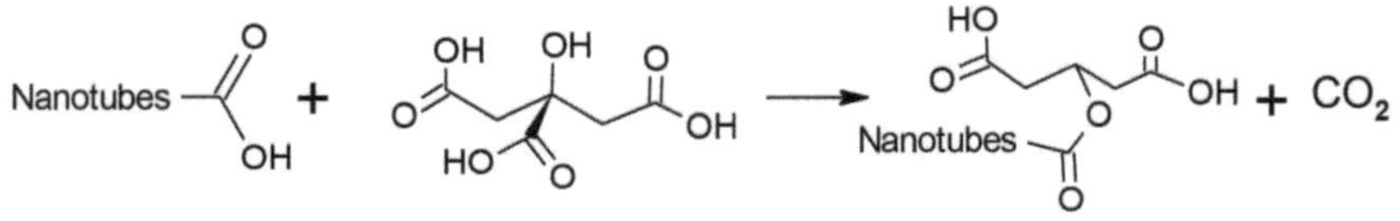

- ***type III masterbatch***: characterized by functionalization of carbon nanotubes with a sulfo-nitric mixture, followed by addition of hydroquinone;

Carbon nanotubes are oxidized in the same way as explained in (a), then a mass m=2.42g of benzene-1,4-diol or hydroquinone: $C_6H_6O_2$ (M=110g/mol) is added.

Given a mole number $n_{hydroquinone}$= (1/16) n H^+ =0.022

Or mole $OH_{hydroquinone}$= mole OH $_{citric\ acid}$ =0.044

The beaker is placed in a water bath containing glycerol and the mixture is heated to 120°C for 3 hours. Esterification of the phenols, even with a reflux set-up and under high pressure, is unlikely, as the rate of esterification is very low, and most of the phenolic hydroxyls remaining in the mixture in excess will participate either in the subsequent reaction with isocyanates in competition with COOH, or in the establishment of hydrogen bonds in competition with the polyol.

The addition of polyether polyols would similarly be mentioned in paragraph (c).

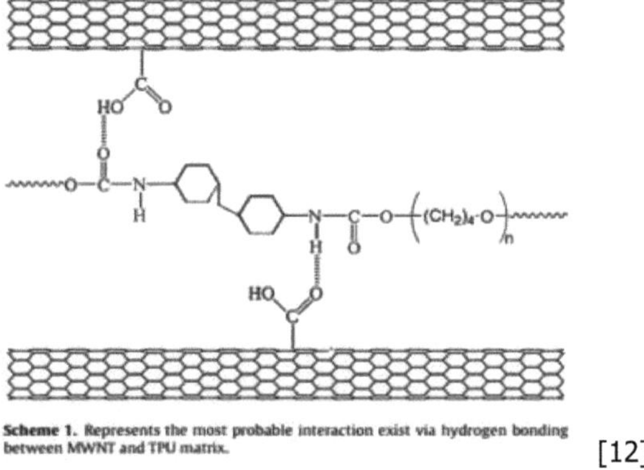

Scheme 1. Represents the most probable interaction exist via hydrogen bonding between MWNT and TPU matrix.

[12]

3.   **Thermogravimetric Analysis** (TGA)

Thermogravimetric analysis was carried out on an instrument (Automatic Multiple Sample Thermogravimetric Analyzer TGA-1000- Navas instruments). Tests were carried out with a constant flow of nitrogen, and at a temperature rise rate of 10°C/min. Sample degradation was monitored over a temperature range from room temperature to 600⁰ C. A 'degradation onset temperature' was defined as the temperature at which the sample had lost 10% of its initial mass.

Method for calculating the semi-degradation temperature

The half-degassing temperature corresponds to the remaining mass of the sample; $m_{1/2}$. This mass is calculated by finding the half-degassed mass $\frac{m_{initiale}-m_{finale}}{2}$, and . $m_{1/2} = 1 - \frac{m_{initiale}-m_{finale}}{2}$.

4.   Tensile Test :

5.   Tensile tests were carried out using two machines; the first was used to cut the specimens to be examined (AC *QUATI- G-VIA-VISMARA-30- ARESE (M) ITALY)*, according to a well-defined shape and dimensions (see FIG-49- a-b- and c).

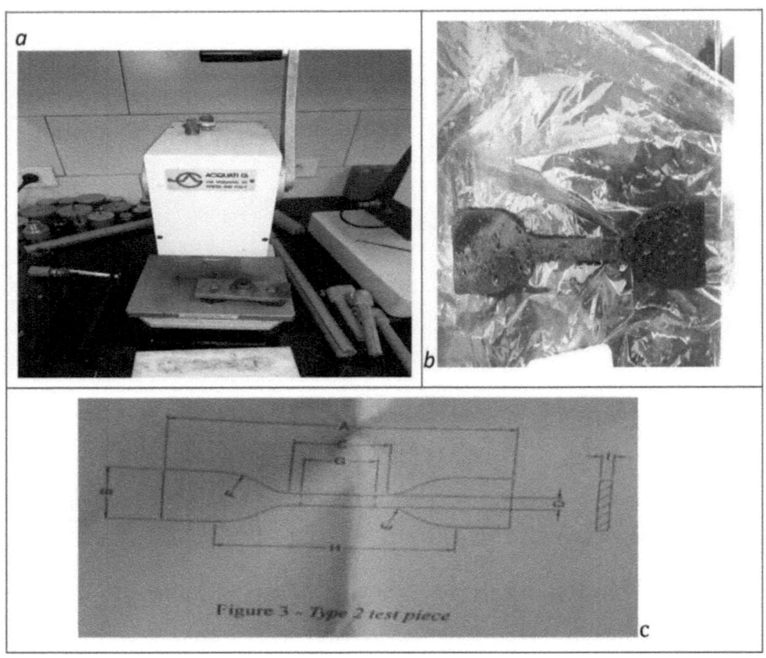

Figure 48-Sample cutting and dimensions

And with its own dimensions as shown in figure (FIG-50) below:

Symbol	Description	Dimensions (mm)
A	Overall length (min.)	115
B	Width of ends	25 ± 1
C	Length of narrow, parallel-sided portion	33 ± 2
D	Width of narrow, parallel-sided portion	$6^{+0.4}_{0}$
E	Small radius	14 ± 1
F	Large radius	25 ± 2
G	Gauge length	25 ± 1
H	Initial distance between grips	80 ± 5
I	Thickness	That of pipe

Table 3 - Dimensions of Type 2 test piece

Figure 49-Sample dimensions

The other, for traction testing, is a specialized machine called (_ACQUATI GIUSEPPE ARESE (MI) A/10IES_), **(Api).**

Tensile testing is carried out at room temperature with an elongation rate of 25mm/minute. From each sample, two specimens are cut and examined, and an average from two values is reported.

The traction testing machine is shown below:

Figure 50 - Tensile test

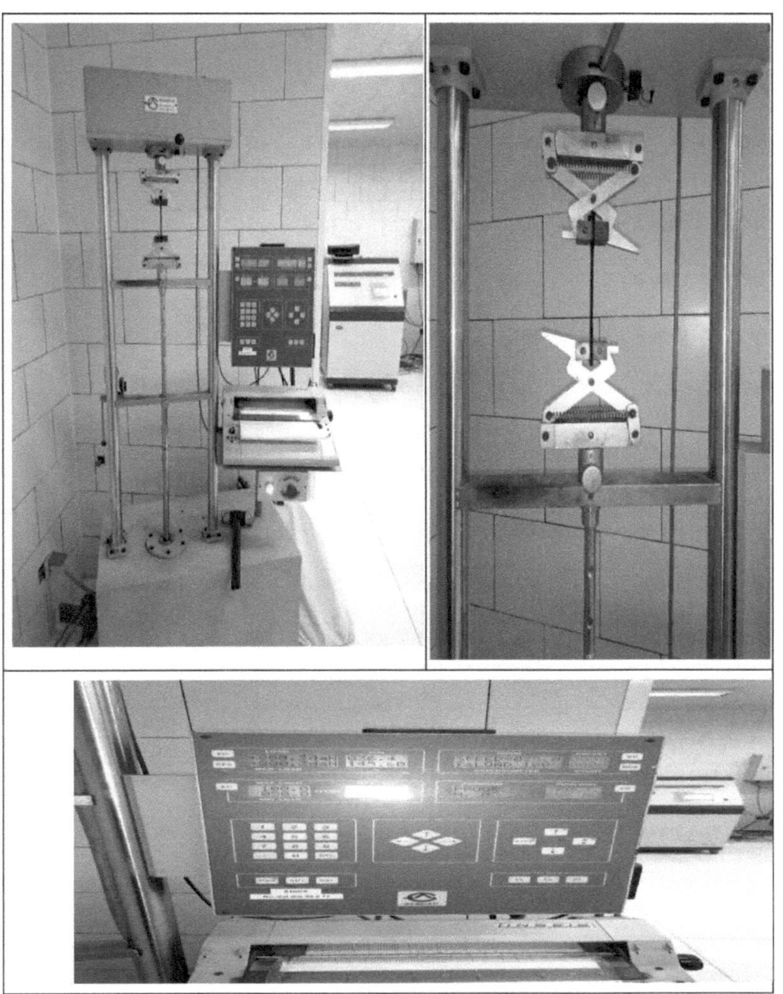

1. Results: Table 14

	Δl 1 in mm	Δl 2 in mm	% elongation 1	Applied force1 inN	% elongation 2	Applied force 2 in N
AN-PU1200-20-NCO(*)-24%-A	170.9	166	683.6	2	664	2
AN-PU1200-18- NCO(*)-24%-A	179.25	207.5	717	2	830	3
AN-PU1200-18- NCO(*)-18%-A	95.6	98.9	382.4	4	395	4
AN-PU1200-18- NCO(*)-12%-A	64.6	59.2	258.4	6	258.4	5
AN-PU1200-18- NCO(*)-6%-A	48.7	53.7	198.8	6	214	7
AN-PU1750-18- NCO(*)-24%-A	43.3	42.7	172.4	10	170.8	10
AN-PU1750-18- NCO(*)-18%-A	85.3	101.6	341.2	6	406.4	8
AN-PU1750-18- NCO(*)-12%-A	89.1	85.7	356.4	8	342.8	8
AN-PU1750-18- NCO(*)-6%-A	76.2	69.4	304.8	11	277.6	10
KC-PU1200-16- NCO(*)-24%-A	279.4	258	1117.6	2	1032	2
KC-PU1200-16- NCO(*)-18%-A	84.9	83.4	339.6	5	333.6	5
KC-PU1200-16- NCO(*)-12%-A	67.4	76.9	269.6	6	307.6	7
KC-PU1200-16- NCO(*)-6%-A	46.2	58.8	184.8	6	235.2	8
KC-PU1750-16- NCO(*)-24%-A	136.6	166.1	546.9	6	664.4	7
KC-PU1750-16- NCO(*)-18%-A	132	122.3	528	7	489	7
KC-PU1750-16- NCO(*)-12%-A	93.3	97.8	373.2	8	391.8	8
KC-PU1750-16- NCO(*)-6%-A	63	73.3	252	10	293.2	11

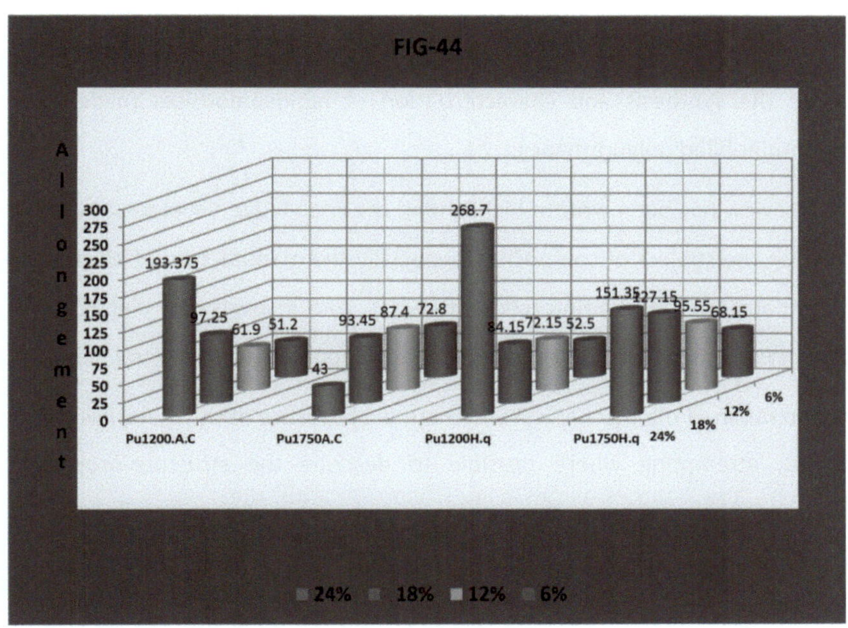

Figure 51-Diagram of variation in elongation with % NC

CONCLUSION

We covered the synthesis and characterization of nanocomposites made from carbon nanotube-filled polyurethanes.

A literature review has been written highlighting the knowledge developed over the last decade on the subject of composites, especially the manufacture of electrically conductive foams, or fibers reinforced with CNTs, and in particular electroactive shape memory polymers, which can have innovative applications in several fields.

We have also outlined the mechanical and thermal properties of the materials under development, attempting where possible to describe the structure-properties relationship of the polymers in question, and especially the influence of incorporating carbon nanotubes into the polymer matrix.

Different composite foams have been prepared and grouped into six differently constituted families. The first two families are obtained with non-functionalized nanotubes, one of which contains a small amount of a Silicon derivative.

The other four families are loaded with two types of oxidized nanotubes: the first type is treated with citric acid after oxidation, while the other type is treated with hydroquinone. The two types of differently functionalized nanotubes are used with two polyols of different molar masses (M=1200, f=3 and M=1750, f=3), to obtain the four families.

Four samples of each family are obtained, containing different mass contents of nanotubes.

Samples were tested for hardness, their homogeneities were examined by scanning electron microscopy (SEM), their stabilities were assessed by ATG thermogravimetric analyses, and their mechanical performance was measured by tensile tests.

The results we have obtained are satisfactory;

1. Hardness

The hardness of a PU-NTC composite is a linear function of the proportion of carbon nanotubes inserted in the matrix. Hardness increases as the percentage of carbons increases, so the presence of Si derivatives reduces hardness and makes the material easier to perforate.

2. Homogeneity

The interactions established in the parent mixture (polyol-24%NTC-COOH/Citric Acid) are strong, and dispersion by simple addition of pure polyol is not evident. Micrographs of all the other composites with hydroquinone show very well dispersed phases, with composites made with molar mass polyol (M=1200gr) showing more homogeneous, continuous textures than those made with molar mass polyol (M=1750gr).

3. Thermal stability

For PUs manufactured with non-functionalized nanotubes, it seems that neither the insertion of Si derivatives nor the percentage of carbon nanotubes present in the PU matrix have any appreciable effect on the temperature at which degradation begins.

samples from four other families show a relationship between the temperature at which degradation begins, on the one hand, and the rate of incorporation of nanotubes into the matrix, on the other; a high rate of incorporation of nanotubes (24%) favors the start of degradation at a low temperature of the order of 275^0 C, this temperature rising steadily as the rate of CNT incorporation decreases, reaching a value of T =308 (or 316) when the percentage of CNT is of the order of 6%.

The temperature at which degradation begins does not characterize PU's thermal stability. The increase in stability can be expressed either by a decrease in the maximum degradation rate, or by the deviation of the maximum point of the degradation peak towards slightly higher temperatures.

Generally speaking, it can be concluded that the introduction of carbon nanotubes increases the thermal stability of the material; in the profiles of the degradation rates of PU samples, the maximum rate decreases gradually with increasing percentages of carbon nanotubes, similarly, the temperatures corresponding to the peak maxima increase with increasing nanotube levels; thus, the degradation rates of Si-containing PUs are clearly greater than those of other polyurethanes, and the insertion of Si decreases their stabilities.

For composites with strong interfacial interactions, each degradation rate profile shows a belly corresponding to a temperature range from {$T=360^0$ to T~395}, in which the degradation rate is average, followed by a maximum rate peak located at a higher temperature. We consider this belly to be **a phase of delayed degradation**, the extent of which is linked to the quantity of interactions between the chains. In our case, these interactions are proportional to the number of NTC-COOH functions. This estimate will be more credible when we notice the relative total or partial absence of this belly in non-functionalized or hydroquinone-treated PU-NTC-composites.

4.   Flame-retardant effect

The mass of residual char at the end of degradation in all samples from all families increased with increasing carbon nanotube content. This observation supports the idea that carbon nanotubes can act as flame retardants.

5.   Mechanical performance

a) For composites made with polyol (M=1200gr), elongation and elongation at break are both functions of the CNT content in the polyol.

composites, the elongation of the composite filled with 24% CNT, relative to the mass of polyol (10% relative to the mass of PU) is of the order of 773%, a behavior close to that of an elastomer, i.e. 3.75 times the elongation of the composite filled with 6% CNT.

b) Fracture strength, expressed as the force applied at the breaking point, is inversely proportional to the percentage of CNTs, with the composite containing the least carbon nanotubes showing the highest fracture strength.

A material that has undergone significant elongation has a lower tensile strength than one that has been slightly elongated.

c) Elongation values are slightly higher for composites treated with hydroquinone, especially the sample containing 24% CNTs which shows elastomer-like elongation (% elongation>1000%), as well as fracture toughness values.

d) For the last family made with a polyol of molar mass (M=1750), a linear relationship can be established between elongation and percentage elongation at break, on the one hand, and percentage loading of functionalized CNTs, on the other.

e) The high fracture toughness values of all the composites correspond to low CNT filler contents, and at the same time to low fracture elongation values, and the converse is true. A comparison of the fracture toughness of all the samples shown in figure (FIG-43-) shows us that the type of nanotube treatment has no influence on the fracture toughness of the composites, but this parameter is dependent on the polyol size.

BIBLIOGRAPHICAL REFERENCES

1.   <u>Effect of nanotube functionalization on the properties of single-walled carbon nanotube/polyurethane composites</u>:

Buffa, F., Abraham, G. A., Grady, B. P., & Resasco, D. (2007). Effect of nanotube functionalization on the properties of single-walled carbon nanotube/polyurethane composites. *Journal of Polymer Science Part B: Polymer Physics*, *45*(4), 490-501.

2.   <u>Synthesis, thermal, mechanical and rheological properties of multiwall carbon nanotube/waterborne polyurethane nanocomposite</u>:

Kuan, H. C., Ma, C. C. M., Chang, W. P., Yuen, S. M., Wu, H. H., & Lee, T. M. (2005). Synthesis, thermal, mechanical and rheological properties of multiwall carbon nanotube/waterborne polyurethane nanocomposite. *Composites Science and Technology*, *65*(11), 17031710.

3.   <u>The thermal and mechanical properties of a polyurethane/multi-walled carbon nanotube composite</u> :

Xiong, J., Zheng, Z., Qin, X., Li, M., Li, H., & Wang, X. (2006). The thermal and mechanical properties of a polyurethane/multi-walled carbon nanotube composite. *Carbon*, *44*(13), 2701-2707.

4.   <u>Deformation-morphology correlations in electrically conductive carbon nanotube-thermoplastic polyurethane nanocomposites</u>:

Koerner, H., Liu, W., Alexander, M., Mirau, P., Dowty, H., & Vaia, R. A.

(2005). Deformation-morphology correlations in electrically conductive carbon nanotube-thermoplastic polyurethane nanocomposites. *Polymer 46*.12(2005) 4405-4420.

5.   **Preparation and characterization of polyurethane-carbon nanotube composites** :

Xia, H., & Song, M. (2005). Preparation and characterization of polyurethane-carbon nanotube composites. *Soft Matter 1*.5(2005): 386-394.

6. **Characterization and mechanical performance comparison of multiwalled carbon nanotube/polyurethane composites fabricated by electrospinning and solution casting**:

Tijing, L. D., Park, C. H., Choi, W. L., Ruelo, M. T. G., Amarjargal, A., Pant, H. R., ... & Kim, C. S. (2013). Characterization and mechanical performance comparison of multiwalled carbon nanotube/polyurethane composites fabricated by electrospinning and solution casting. *Composites Part B: Engineering*, *44*(1), 613-619.

7. **Carbon nanotube-reinforced polyurethane composite fibers:** Chen, W., Tao, X., & Liu, Y. (2006). Carbon nanotube-reinforced polyurethane composite fibers. *Composites Science and Technology*, *66*(15), 3029-3034.

8. **Optically Active Multi-Walled Carbon Nanotubes for Transparent,Conductive Memory-Shape Polyurethane Film**

Jung YC, Kim HH, Kim YA, Kim JH, Cho JW, Endo M, et al. Optically active multiwalled carbon nanotubes for transparent, conductive memory-shape polyurethane film. Macromolecules 2010;43:6106- 12

9. **Carbon nanotube-polymer composites: chemistry, processing, mechanical and electrical properties :**

Spitalsky, Z., Tasis, D., Papagelis, K., & Galiotis, C. (2010). Carbon nanotube-polymer composites: chemistry, processing, mechanical and electrical properties. *Progress in polymer science*, *35*(3), 357401.

10. **Electroactive shape-memory polyurethane composites incorporating carbon nanotubes:**

Cho, J. W., Kim, J. W., Jung, Y. C., & Goo, N. S. (2005). Electroactive shape-memory polyurethane composites incorporating carbon nanotubes. *Macromolecular Rapid Communications*, *26*(5), 412-416.

11. Preparation and characterization of conductive carbon nanotube-polyurethane foam composites :

You, K. M., Park, S. S., Lee, C. S., Kim, J. M., Park, G. P., & Kim, W. N. (2011). Preparation and characterization of conductive carbon nanotube-polyurethane foam composites. *Journal of materials science*, *46*(21), 6850-6855.

12. Preparation and properties of acid-treated multiwalled carbon nanotube/waterborne polyurethane nanocomposites:

Kwon, J. Y., & Kim, H. D. (2005). Preparation and properties of acid-treated multiwalled carbon nanotube/waterborne polyurethane nanocomposites. *Journal of Applied Polymer Science*, *96*(2), 595-604.

13. Conductive network formation in the melt of carbon nanotube/thermoplastic polyurethane composite:

Zhang, R., Dowden, A., Deng, H., Baxendale, M., & Peijs, T. (2009). Conductive network formation in the melt of carbon nanotube/thermoplastic polyurethane composite. *Composites Science and Technology*, *69*(10), 1499-1504.

14. Biocompatible hyperbranched polyurethane/multi- walled carbon nanotube composites as shape memory materials:

Deka, H., Karak, N., Kalita, R. D., & Buragohain, A. K. (2010). Biocompatible hyperbranched polyurethane/multi-walled carbon nanotube composites as shape memory materials. *Carbon*, *48*(7), 2013-2022.

15. Polyurethane-Carbon nanotube nanocomposites prepared by in-situ polymerization with electroactive shape memory:

Jin Yoo, H., Chae Jung, Y., Gopal Sahoo, N., & Whan Cho, J. (2006).

Polyurethane-Carbon nanotube nanocomposites prepared by in-situ polymerization with electroactive shape memory. *Journal of Macromolecular Science, Part B*, *45*(4), 441-451.

16. **Shape-memory polyurethane/multiwalled carbon nanotube fibers:**

Meng, Q., Hu, J., & Zhu, Y. (2007). Shape-memory polyurethane/multiwalled carbon nanotube fibers. *Journal of Applied Polymer Science, 106*(2), 837-848.

17. **Mechanical and thermal properties of functionalized multiwalled carbon nanotubes and multiwalled carbon nanotube-polyurethane composites:**

Chen, X., Wang, J., Zou, J., Wu, X., Chen, X., & Xue, F. (2009). Mechanical and thermal properties of functionalized multiwalled carbon nanotubes and multiwalled carbon nanotube-polyurethane composites. *Journal of applied polymer science, 114*(6), 3407-3413.

18. **Electroactive shape memory performance of polyurethane composite having homogeneously dispersed and covalently crosslinked carbon nanotubes:**

Jung, Y. C., Yoo, H. J., Kim, Y. A., Cho, J. W., & Endo, M. (2010).

Electroactive shape memory performance of polyurethane composite having homogeneously dispersed and covalently crosslinked carbon nanotubes. *Carbon, 48*(5), 1598-1603.

19. **Thermosetting polyurethane multiwalled carbon nanotube composites:**

McClory, C., McNally, T., Brennan, G. P., & Erskine, J. (2007). Thermosetting polyurethane multiwalled carbon nanotube composites. *Journal of Applied Polymer Science, 105*(3), 1003-1011.

20. **Carbon nanotube-polyurethane shape memory nanocomposites with low trigger temperature:**

Gu, S., Yan, B., Liu, L., & Ren, J. (2013). Carbon nanotube- polyurethane shape memory nanocomposites with low trigger temperature. *European Polymer Journal, 49*(12), 3867-3877.

21. **Preparation, characterization and properties of acid functionalized multi-walled carbon nanotube reinforced thermoplastic polyurethane nanocomposites:**

Barick, A. K., & Tripathy, D. K. (2011). Preparation, characterization and properties of acid functionalized multi-walled carbon nanotube reinforced thermoplastic polyurethane nanocomposites. *Materials Science and Engineering: B*, *176*(18), 1435-1447.

22. **Electrical conductivity and major mechanical and thermal properties of carbon nanotube-filled polyurethane foams:**

Yan, D. X., Dai, K., Xiang, Z. D., Li, Z. M., Ji, X., & Zhang, W. Q. (2011).

Electrical conductivity and major mechanical and thermal properties of carbon nanotube-filled polyurethane foams. *Journal of applied polymer science*, *120*(5), 3014-3019.

23. **Influence of hard segment content and nature on polyurethane /multiwalled carbon nanotube composites:**

Fernândez-d'Arlas, B., Khan, U., Rueda, L., Coleman, J. N., Mondragon, I., Corcuera, M. A., & Eceiza, A. (2011). Influence of hard segment content and nature on polyurethane/multiwalled carbon nanotube composites. *Composites Science and Technology*, *71*(8), 1030-1038.

24. **Study of electroactive shape memory polyurethanecarbon nanotube hybrids:**

Lee, H. F., & Yu, H. H. (2011). Study of electroactive shape memory polyurethane-carbon nanotube hybrids. *Soft Matter*, *7*(8), 38013807.

25. **Effect of functionalized carbon nanotubes on molecular interaction and properties of polyurethane composites:**

Sahoo, N. G., Jung, Y. C., Yoo, H. J., & Cho, J. W. (2006). Effect of functionalized carbon nanotubes on molecular interaction and properties of polyurethane composites. *Macromolecular chemistry and physics*, *207*(19), 1773-1780.

26. **A flexible multifunctional sensor based on carbon nanotube/polyurethane composite:**

Slobodian, P., Riha, P., Benlikaya, R., Svoboda, P., & Petras, D. (2013). A flexible multifunctional sensor based on carbon

nanotube/polyurethane composite. *IEEE Sensors Journal*, *13*(10), 4045-4048.

27. **Electroactive shape memory effect of polyurethane composites filled with carbon nanotubes and conducting polymer:**

Sahoo, N. G., Jung, Y. C., & Cho, J. W. (2007). Electroactive shape memory effect of polyurethane composites filled with carbon nanotubes and conducting polymer. *Materials and Manufacturing Processes*, *22*(4), 419-423.

28. **Shape memory effect and mechanical properties of carbon nanotube/shape memory polymer nanocomposites:** Ni, Q. Q., Zhang, C. S., Fu, Y., Dai, G., & Kimura, T. (2007). Shape memory effect and mechanical properties of carbon nanotube/shape memory polymer nanocomposites. *Composite Structures*, *81*(2), 176-184.

29. **Shape memory properties of multi-walled carbon nanotube /polyurethane composites prepared by in situ polymerization:**

Bai, Y., Zhang, Y., Wang, Q., & Wang, T. (2013). Shape memory properties of multi-walled carbon nanotube/polyurethane composites prepared by in situ polymerization. *Journal of Materials Science*, *48*(5), 2207-2214.

30. **Preparation and characterization of polyurethane/multiwalled carbon nanotube composites:**

Guo, S., Zhang, C., Wang, W., Liu, T., Tjiu, W. C., He, C., & Zhang, W. D.

(2008). Preparation and characterization of polyurethane/multiwalled carbon nanotube composites. *Polymers & Polymer Composites*, *16*(8), 501.

31. Study of electroactive shape memory polyurethanecarbon nanotube hybrids:

Lee, H. F., & Yu, H. H. (2011). Study of electroactive shape memory polyurethane-carbon nanotube hybrids. *Soft Matter, 7*(8), 38013807.

Printed by Books on Demand GmbH, Norderstedt / Germany